"Catherine Rakow's present work fills a void on the progress and thinking of Dr. Bowen before his NIMH Family Study Project. Her work is a seminal contribution towards the understanding of Bowen's development of his theory and applications of family psychotherapy."

John F. Butler, *Rose Street Mental Health Care, Wichita Falls, Texas*

"Rakow has cast a broad net, using extensive archival materials to document Bowen's process at arriving at a theory, which is much broader than his 8 concepts. A theory based on factual clinical observations that led to "keeping the transference in the family", "non-mothering attitude", functions of symptoms. The book will challenge the reader's assumptions and principles of one's life and work."

Clarence Boyd, *editor, Commitment to Principles, the letters of Murray Bowen, M.D.*

"An important contribution to the foundation of facts of how Bowen theory developed over time and clues for where development can continue. Consistent with Bowen theory, this work does little to tell the reader how or what to think but invites curiosity and provokes theoretical inquiry."

Amie Post, *Faculty, The Bowen Center for the Study of the Family, Washington DC*

"Ms. Rakow has not only captured an important period in Bowen's work, but she offers us an approach that we each might take up in furthering our understanding of our various human systems. In that spirit, her book is essential reading for marriage and family therapists, mental health professionals, organizational consultants, community activists, members of communities of faith, students of human behavior, family medicine practitioners, and those interested in the history of medicine. Catherine Rakow has been a dedicated scholar of the Murray Bowen Archives for many years. She has steadfastly examined the detailed evidence of those earlier years of Dr. Bowen's life and work as a medical doctor, psychiatrist, and researcher when he was developing his theory of human functioning. In time this effort became his natural family systems theory, known today as Bowen Family Systems Theory or Bowen theory. His approach to thinking and to researching oneself in one's human system is now taught and applied worldwide. It has been my privilege to teach this material in my current role as a faculty member at the Western Pennsylvania Family Center, (as is Ms. Rakow), and, to observe the remarkable changes in functioning of so many of the learners once they contact and research their own family, organizational, and community systems—once they make sense of their own human lives."

N. Michel Landaiche III *is WPFC Faculty Member and Human Relations Consultant; he has authored* Groups in Transactional Analysis, Object Relations, and Family Systems: Studying Ourselves in Collective Life

MAKING SENSE OF HUMAN LIFE

Drawing from rich archival material, this book provides unprecedented access to the professional documents and historical context surrounding the life's work of Dr. Murray Bowen (1913–1990), medical doctor, psychiatrist and pioneering researcher of Family Systems Theory.

To understand the origins and evolution of this theory, Catherine Rakow explores Bowen's early years as a psychiatrist at the Menninger Foundation—at which time he became curious about the possibility of determining a factual basis for psychoanalytic theory—and explains how this research would foreground Bowen's lifelong study of the family unit at the National Institute of Mental Health. From those seminal years of study and observation, Rakow explains how Bowen developed Family Systems Theory: A theory of human functioning that conceives of family as a naturally occurring, regenerating system. Rakow's close engagement with Bowen's practice and influences at this time allows for a fulsome account of the research process that Bowen undertook to develop this innovatory approach.

In this book, Rakow demonstrates the value of Bowen's work as a model and research methodology for those exploring the role of theory in improving family relationships, making it essential reading for marriage and family therapists, mental health professionals, students, those interested in the history of medicine, and curious individuals alike.

Catherine M. Rakow is an independent researcher, has spent nearly 30 years reviewing Murray Bowen archival materials at the National Library of Medicine, History of Medicine Division in Bethesda, MD and the Bowen family collection in Williamsburg, VA.

MAKING SENSE OF HUMAN LIFE

Murray Bowen's Determined Effort Toward Family Systems Theory

Catherine M. Rakow

Routledge
Taylor & Francis Group

NEW YORK AND LONDON

Cover image: © uncle-rico, Getty Images

First published 2023
by Routledge
605 Third Avenue, New York, NY 10158

and by Routledge
4 Park Square, Milton Park, Abingdon, Oxon, OX14 4RN

Routledge is an imprint of the Taylor & Francis Group, an informa business

© 2023 Catherine M. Rakow

Every effort has been made to contact copyright-holders. Please advise the publisher
of any errors or omissions, and these will be corrected in subsequent editions.

Library of Congress Cataloging-in-Publication Data
Names: Rakow, Catherine M., author.
Title: Making sense of human life: Murray Bowen's determined effort toward
family systems theory/Catherine M. Rakow.
Description: New York, NY: Routledge, 2022. |
Includes bibliographical references and index. |
Identifiers: LCCN 2022004562 (print) | LCCN 2022004563 (ebook) |
ISBN 9780367461522 (hardback) | ISBN 9780367461546 (paperback) |
ISBN 9781003027287 (ebook)
Subjects: LCSH: Bowen, Murray, 1913–1990. | Psychiatrists–United States–Biography. |
Physicians–United States–Biography. | Family psychotherapy.
Classification: LCC RC339.52.B69 R35 2022 (print) |
LCC RC339.52.B69 (ebook) | DDC 616.89/156–dc23/eng/20220510
LC record available at https://lccn.loc.gov/2022004562
LC ebook record available at https://lccn.loc.gov/2022004563

ISBN: 978-0-367-46152-2 (hbk)
ISBN: 978-0-367-46154-6 (pbk)
ISBN: 978-1-003-02728-7 (ebk)

DOI: 10.4324/9781003027287

Typeset in Bembo
by Newgen Publishing UK

To Paul, for 59 years my partner in this natural experiment called family.
The bonds were constant.

CONTENTS

FOREWORD

I am pleased and honored to write a Foreword to Catherine Rakow's *Making Sense of Human Life: Murray Bowen's Determined Effort Toward Family Systems Theory*, a book that draws from Murray Bowen's archival legacy to introduce to the world how he worked and how he developed his theory, his "gift to the world." Until now, the mental health profession has known about Murray Bowen's professional odyssey toward a science of human behavior primarily through his presentations, workshops, videos of clinical work and teaching, and the Georgetown Family Center and Bowen Center's long-standing symposiums and training program. Published works include papers in edited volumes on family therapy and *Family Therapy in Clinical Practice*, his volume of papers he considered his most important. Toward the end of his life, he published his odyssey in *Family Evaluation* (Kerr & Bowen, 1988).

During the 1990s, when his wife, LeRoy Bowen, gave researchers limited access to his National Institute of Mental Health papers, Catherine Rakow, Monika Baege (2005) and Deborah Weinstein (2013) conducted research. In 2002, after the Bowen Family had gifted his legacy to the National Library of Medicine (NLM), Clarence Boyd (2016) and Jack Butler (2013) researched before NLM closed the collection in 2007, pending resolution of Personal Health Information requirements. To assist NLM's goal to open the collection to researchers, the Bowen family gave NLM permission to invite Rakow to identify records with information that could be known only in the archives. Not one to pass on an opportunity, she took on the monumental task of reviewing over 250 linear feet of unpublished papers, early drafts, clinical notes and correspondence. As she worked, Rakow learned the scope of Bowen's archival legacy. In 2011, when the Bowen family moved early records and family correspondence to Williamsburg, where I live, we invited Rakow to oversee this project. With the assistance of Bowen family members, The Murray Bowen Archives Project volunteers worked under Rakow's guidance, and she continued her research. Taken as a whole, these sources reveal his genius, namely his ability to remain neutral amid emotional trauma, to distinguish between emotion and fact, and to observe broadly and deeply to see interpretations based on emotion, not fact. Drawing this ability to observe, he conducted scientific research, much like Charles Darwin, observing life inductively to differentiate between emotion, interpretation and fact. By weighing facts against then-current Freudian theory, he could deduce his way from focusing on the individual, conceiving the family as a unit and eventually developing a theory of an emotional system as a part of nature.

My siblings and I have understood our father's deep commitment to making a difference and solving unsolvable problems. As a child, I heard him tell a story that has become common knowledge among his students. We learned that his family, who owned a funeral home, used their hearse to transport patients from his hometown of Waverly, Tennessee, to the nearest hospital in Nashville. One trip involved driving a young teenage girl who had taken sick early in the morning. He witnessed doctors' struggle, not knowing how to treat her, and by evening she died. What did he focus on, the girl struggling for life? No, he focused on the doctors and the medical system. Afterward, he vowed to become a physician and solve problems medicine could not.

I learned Bowen theory in my home and by attending professional meetings. However, it took years for me to appreciate how my father's astute observational and analytical capabilities, combined with his ability to be neutral and stay outside tense emotional situations, enabled him to become the ethologist who observed life and all relationships between humans and their natural world. Going further, one could say his motivation to go where none had traveled and his ability to live in emotional environments and see functional facts enabled him to discover how the family formed an emotional UNIT. This knowledge also enabled him to recognize differentiation as the cornerstone of the emotional system that grounds the behavior of humans and other life-forms.

When asked how he developed his theory, he would recall that his involvement on the library committee permitted him to read widely in the sciences. When pressed in his later life, he would say, "I dreamed it up." But there was more to it. I understood he read widely searching for scientific principles, specifically evolutionary theory, that placed humans and their families in nature. I heard him say that his observations of human emotions did not support the theory he was learning and that after he released his patients to their families, they would regress and return with the same symptoms. He would conclude by emphasizing how patients were symptoms; the problem lay within the family.

As a researcher with a long career working with documentary, archaeological, ethnographic and scientific data, I am in awe of Rakow's work. She has read widely, researched deeply and broadly in Bowen's archives, synthesized information and embedded Bowen into the professional context in which he worked. The outcome of this effort has the potential to brush away preconceptions and open the door to reveal Bowen at work. You will not be disappointed—her work will become a classic.

Rakow's research, however, shows that Bowen's odyssey toward science began as a child. Born in Waverly, Tennessee, a small town 60 miles west of Nashville, Bowen lived on a farm that produced virtually all their own food. He often spoke of the importance of his childhood and what he learned from his father. Jess Sewell Bowen, a farmer, hunter, town leader and part-owner of a business, taught his son to appreciate the land, nature and all wildlife. Jess Bowen often remarked how he either knew a person personally or knew their kin by how they walked. He taught by action how to know the importance of the community in which they lived. Like his father, Murray Bowen was an observer who understood the vital role social relationships play in sustaining the very fabric of an agricultural community. Bowen's genius, however, was far greater—he had an intuitive understanding of the role emotions play in EVERY relationship—within and amongst families, communities and entire nations. He could see the emotions undergirding well-thought-out positions when others only saw the logic. He often remarked, "Humans can think about, think about, think about." To him, what counted was not what humans said but what they did, and he observed behavior as would a biologically trained ethologist.

While the profession generally assumes Bowen's interest in psychiatry emerged while serving on the war front, Rakow demonstrates that his interest emerged earlier, while he worked in Fort Bragg's psychiatric ward under Chief of Psychiatry, Dr. Norman Brill. A poem written by a private at Fort Bragg shows Bowen left a mark on Ward 98. Even as a young physician, he could relate to the intensity of emotions, keep himself outside of the emotional arena and at the same time keep individuals accountable for their actions. Written in 1943 shortly before Bowen shipped out to the European war front, this private's poem encapsulates Bowen's innate strength from his upbringing and ability to stand above the pull of emotional situations.

TO A GREAT GUY WHO IS TOO SWELL TO EVER BE A GREAT OFFICER

This to you, my Captain, I give in parting token
Of the kindness you have shown me, of the faith that I have broken,
I've met a lot of officers that caused me grief and woe,
But you are about the first one that I've ever liked to know.

I know of all the hell that I have raised, my wide and sneering mouth,
I guess that I'm a disgrace to all that's in the South,
I wish that I were Emerson and rich in eloquent phrase,
That I might deeply pay you tribute for your kind and gracious ways.

Sometimes I sit and curse you, I solemnly will confess,
For I think you're the guy that got me in this mess,
But that cursing doesn't last long, for I can't long erase
The vision of your tireless hands, and smiling pleasant face.

I've seen you work from dawn to night, and then go work so more,
I'll swear the guy is crazy, that office he adores
In silent worship all of us, would like for you to stay
And many a madman's heart is sick because you're going away.

Even now I hear you pecking, in your old two finger style,
Patient and enduring, and sometimes you stop to smile,
Occasionally you stop to swear, and fairly pull your hair,
But since this is the Army, I can see why you should (c)are.

The Army pulled me under, and tore up this ugly head,
But you had enough of it to drive a man plumb dead,
Still you are not complaining, though I think you must be sad.
You've had more than I have had and here it's driven me mad.

We fear the man what takes your place, he can't come up to you,
For not many men are Bowens, God made a mighty few,
So if you please take these lines from old Ward 98

As profound appreciation, from those who sit and wait.
Pvt. (xx).

By mining Murray Bowen's archives, Rakow reconstructs his pursuit of the science of mental health and how he moved from an individual focus to placing individual's symptoms into the family in which they belong. Her evidence shows that within a year of Bowen's arrival at the Menninger Foundation he had discovered the reality of the nuclear family interface and had established the foundation of what would become known as Bowen theory by the time he left Topeka. Her text is rich with information about when and how this occurred.

Rakow's greatest gift to the future of Bowen theory is her revealing HOW Murray Bowen approached problems in search of the unknown and how he withstood the psychoanalytic community's often severe criticism for embracing new facts as they emerged. By drawing on a draft of Bowen's Epilogue (Kerr & Bowen, 1988), Rakow lays out his approach, one that current and future mental health professionals and other researchers attempting to discover the unknown would be well advised to absorb and emulate. The future of Bowen theory rests on researchers embracing the inductive principles that grounded Murray Bowen's odyssey.

Bowen's innate curiosity led him to wonder about the harmony he saw in nature, humans and all life-forms from his youth. As an aspiring psychiatrist in training at the Menninger Foundation (which promoted creativity), Bowen found himself free to explore new ideas. Moving as fast as humanly possible, he explored current literature to discover connections between evolution and the diversity present in human behavior. Finding Darwin's understanding of evolution and science to be most compatible, he embraced the scientific inductive approach that Darwin described as "groupings of facts so that general laws or conclusions may be drawn from them" (Darwin, 1887, 48). In a draft of his Epilogue in Family Evaluation (1988), Bowen laid out his approach to reading a wide body of literature, assessing the many variables he encountered, and evaluating and recognizing patterns before forming tentative hypotheses and resolving discrepancies. His path took years, starting from his understanding of evolution and progressing to observing and cross-checking books he read everything with he learned across his coursework, clinical practice and communications with patients, family and staff. An important goal was to separate feeling-subjective states from the objective, non-feeling states, which he did by identifying and eliminating statements based on feelings, imagination, tuition, mythology or any form of subjectivity. Gradually, by weighing his observations against Freudian theory, finding discrepancies and integrating the knowledge he accumulated through reading scientific literature, he could refine his understanding of the reciprocity in family interactions (Bowen, 1986a).

I am in awe of what Bowen accomplished in a remarkably short time; how he integrated his academic and clinical training to make a fundamental shift from individually based psychoanalysis that separated patients from their families to integrating treatment of patients and their families. Rakow's analysis clarifies how the development of Bowen theory could have occurred ONLY in an institutional setting that combined academic research with therapeutic techniques in what amounted to a laboratory where he could observe the behavior of the clinical staff, the patients and their families. In this work environment, Bowen could evaluate the objective status of his findings and then rethink and regroup with another experiment to implement his findings. In this way, he could modify how mental health workers relate to patients and their families, developing new policies and therapies that would leave emotions with the family, and enriching how families and their symptomatic members could find ways to resolve their problems. How Bowen managed with himself in the middle is a lesson for anyone working on their differentiation.

His accomplishments in shifting from individual to family therapy are too many to reference; what will suffice here is to relay how Murray Bowen planted and nurtured foundational principles during his time at the Menninger Foundation. In Freudian analysis, the patient/psychiatrist relationship was sacrosanct, and the nursing staff served as the patient's parent, providing for their every need. Social workers managed the families, a policy that kept families separate and less able to find solutions and take actions of their own accord.

Bowen initiated change by modifying how the clinical staff worked with patients, that is, not taking sides and not giving advice. And by promoting and integrating communications between staff, patients and families, Bowen broke open the patient/psychiatrist and family/social worker silos. By requiring staff to remain neutral, families were better able to take ownership of their emotional issues.

While navigating institutional regulations protecting the patient/therapist relationship, Bowen sought ways to interact with families by way of using the mail, then engaging through social services, interacting with them during off-hours and communicating with them during periodic visits. Finally, when he worked as a director in an out-patient clinic, he was freer to explore and discover a model where families would have responsibility for their problems. To use Rakow's words, Bowen wove together his institutional and out-patient efforts to weave the "warp and woof" of his future and his theory that emerged out of a lifetime of continued observation of human life in its emotional expression (Rakow text, Chapter 2).

Equally remarkable is how Bowen modified existing therapies such as anaclitic or regression therapy. In contemporary psychoanalysis, the goal of the treatment was to resolve significant past events believed to interfere with a person's present mental and emotional wellness. Instead, Bowen chose to focus on the life-giving, natural closeness between a mother and her offspring and what he believed could activate an inner life force within self to manage the environment and make decisions for themselves. Nurses were expected to provide for their patients' little and big desires with no advice and to stay OUT of the family's emotional field. This step permitted the family to contain emotions within their relationships. Then, when the patient regressed to a preverbal level, they could activate their resources within self to manage the environment on their own. Bowen reported success and afterward offered supportive psychotherapy combining social work meetings with the family, all aimed at increasing the person's level of awareness of competence and ability to make their own decision. Here is the core of what would emerge as Bowen theory.

Rakow's work is clear; Bowen worked methodically, moving as fast as he could within the environment he worked. His genius lay in his ability to stay focused on his core belief that humans were part of nature and to withstand the reactions to his challenges. Knowing the boundaries and the benefit of expanding his thinking within them, he was able to stay out of the emotional forces existing within institutions, staff, patients and families to develop a better understanding of the human condition and to develop therapeutic techniques that could assist individuals and their families in solving their own problems. I encourage all to immerse themselves into her text, as it reveals a genius at work.

Joanne Bowen, Ph.D.
Former Senior Curator of Environmental Collections, Colonial
Williamsburg Foundation and Research Professor in Anthropology,
The College of William and Mary
Founding President and Current Board Member, Bowen Family
Representative to the Murray Bowen Archives Project

References

Baege, M. (2005). *Family process influences on the resilient responses of youth.* [Doctoral dissertation, University of Vermont] Graduate College Dissertations and Theses. 1079. https://scholarworks.uvm.edu/graddis/1079.

Bowen, M. (1986a). [Draft]. (Acc. 2006-003, Box 8, Folder Drafts of 'odyssey'—In theoretical principle).

Boyd, C. (2016). *Commitment to principles*. Retrieved November 20, 2021, from The Murray Bowen Archives Project: http://murraybowenarchives.org/boyd-book.

Butler, J. F. (Ed.). (2013). *The origins of family psychotherapy: The NIMH family study project*. New York: Jason Aronson.

Darwin, C. (1887). Autobiography. In F. Darwin (Ed.), *The life and letters of Charles Darwin: Including an autobiographical chapter*. John Murray. https://doi.org/10.5962/bhl.title.1416.

Weinstein, D. F. (2013). *The pathological family: Postwar America and the rise of family therapy*. Ithaca, NY: Cornell University Press.

PREFACE

It began with simple curiosity.

In June 1993, I was searching for a 1957 conference paper referenced by Dr. Murray Bowen in his 1978 book, *Family Therapy in Clinical Practice*. That paper was intriguingly titled "The 'Action Dialogue' in an Intense Relationship: The Study of a Schizophrenic Girl and Her Mother." I thought it might offer something useful for my relationship with my mother. Written by Robert H. Dysinger, the paper recorded his effort to be a more factual observer while part of Bowen's National Institute of Mental Health (NIMH) research project. He described efforts to control his opinions, judgments and preconceived interpretations in order just to observe what he called "gross action" (Dysinger, 2003, 121). His intent was to capture the interactional patterns between a mother and daughter without focusing on the content of the words exchanged within that family unit.

To track the paper down, I made inquiries at what was then called the Georgetown Family Center (GFC), founded by Dr. Bowen, where I was completing postgraduate studies in Bowen's natural family systems theory. I learned that Bowen had saved all the materials from his 1954–1959 research project at NIMH, moving them to his home in Bethesda, MD, several years before his death, in 1990. That captured my interest, and I wanted to see those materials. So I submitted a proposal to his widow, explaining my interest.

LeRoy Bowen generously responded by suggesting she loan her husband's NIMH papers to the library at GFC. So I loaded six file drawers into my Volkswagen Rabbit to drive them from Bethesda to the library at GFC, in Washington, DC. In those drawers were over 200 folders in five general categories: (1) drafts of papers, some handwritten; (2) clinical records; (3) administrative notes, including meeting notes, correspondence and memos; (4) published materials such as newsletters, magazines and newspaper articles and (5) extensive correspondence between

Dr. Bowen and other professionals and with his clients. I eventually detailed these contents in my first article about Bowen's archive (Rakow, 1994).

From November 1993 to June 2006, I studied and wrote about the archival records then at GFC. In return for the space to conduct my work and, at Dr. Michael Kerr's suggestion, I wrote a synopsis of materials as I reviewed them. These efforts were recognized when I received the Polly Caskie award in 2004 for cataloging the records of the NIMH research project and adding significantly to the history of the development of Bowen's groundbreaking theory.

In 2006, the Bowen family donated the archival materials to the National Library of Medicine (NLM). When the materials transferred to NLM from GFC (at that point renamed The Bowen Center for the Study of the Family), I also transferred my research activities to NLM. Each month I traveled from my home in western Pennsylvania, either to the Bowen Center in Washington, DC, or to NLM in Bethesda, MD. During that time, I also reviewed additional records located at NLM covering Dr. Bowen's professional life from 1960 until his death, in 1990. In return for permission by NLM to continue my work, I created a finding aid to the NIMH portion of the Bowen archives. It is currently available to other researchers. This aid includes documentation on where to find information on the research families, as well as a reference to Bowen's drafts and manuscripts from that era, many of which are handwritten and untitled. I concluded that study in 2008.

That same year, a group of Bowen-trained professionals under Dr. Joanne Bowen's direction formed what is now known as The Murray Bowen Archives Project. Its purpose is to assist with preserving and processing the records at NLM. Under this group's auspices, I continued to work with the archives at NLM until 2012. Then I took on the role of project manager in an effort to organize the Bowen family's private documents and to separate personal papers from the professional papers. This set of records is comparable in volume to Bowen's professional papers already at NLM. Those additional materials then also became part of the Bowen family's deed of gift to NLM, though, as of this writing, they have not yet transferred there.

I had written several introductions to archival materials published in the *Family Systems Journal* and had contributed a page on research to the Bowen Center's website. My first published review of the materials, "Analyzing Observational Data from Bowen's NIMH Project: Two Months of Nursing Notes: August 1955 and October 1956," appeared in *The Family Systems Forum*, a journal of the Center for the Study of Natural Systems and the Family, in Houston, TX, in 2013.

Many times along the way, someone asked me about writing a book. Through all these years of reviewing the materials, I kept notes on the documents that illustrated Bowen's development of a new theory of human behavior based in science. Though I had created a timeline of events, I considered important in the development of Bowen's theory, I did not want to simply chronicle those advances linearly. Gradually, however, I became curious about a comment in Bowen's, "Epilogue: An Odyssey toward Science," his chapter in *Family Evaluation* (Kerr & Bowen, 1988). A conversation about it with the book's co-author, Dr. Michael Kerr, in 2011, brought the research question for this book into my head. In that conversation, I mentioned that in Bowen's chapter, Bowen said he had created a new theory while he was still in Kansas, long before coming to NIMH. Dr. Kerr replied he had also noticed that comment. This conversation fit well with my original interests from eighteen years earlier. What were the ideas and observations forming the basis of the theory and how did these come together to form a new theory? When did he speak about this new theory to others?

I hope my findings in this book stimulate further interest in the NLM archives, in Bowen's seminal research, and in the later evolution of his theory from 1955 to 1990. While the 1954–1959 NIMH research project is well known among clinicians and students familiar with Bowen's work, the preceding decade of his research is less known. The inner workings of Bowen's NIMH research project offer a fertile area for exploration. Even Bowen referred to his NIMH project as mainly documenting the shift from a focus on schizophrenia to laying the groundwork for a more expansive systems theory. As Bowen noted, the extraordinary wealth of details from that period of research

> was really never written up well because it opened so many doors to new areas, and it was so quickly incorporated into the new areas. There was no reason to ever document it, in writing or in film, as a thing unto itself.
>
> *Bowen, 1978*

Further clues about his thoughts on human functioning may yet remain in that earlier decade's collection of papers. Perhaps this book's effort can be a springboard to further explorations of the archival wealth that resides at the NLM.

I also feel a sense of responsibility to make Bowen's theory available more widely. It is often said that Bowen "gave his theory to the world" in 1988. Being one of many who accepted that gift, I likewise wanted to make something out of my own years of reviewing, reflecting, presenting and writing articles. I wanted to see what I'm made of by incorporating the theory beyond my family relationships. Perhaps this work of mine will have interest enough to add to the historical record. I would like to leave that legacy. There are still many details left to be filled in by others.

I believe that the formulation and advancement of Bowen's natural systems theory is part of a paradigm shift. There is a broader process of moving evolutionary theory toward including the human as part of all living things. For me, this has the larger implication of linking all living things to functioning systems in nature, which would include our human families.

References

Bowen, M. (1978). [Letter to a chaplain on January 3, 1978]. L. Murray Bowen papers, National Library of Medicine, History of Medicine Division. (Acc. 2007-012, Box 3, Folder MCV), Bethesda, MD.

Dysinger, R. (2003). The 'Action Dialogue' in an intense relationship: The study of a schizo-phrenic girl and her mother. *Family Systems Journal, 6*(2), 117–120.

Kerr, M. E., & Bowen, M. (1988). *Family evaluation: An approach based on Bowen theory.* New York: W. W. Norton.

Rakow, C. (1994). Bowen archives the NIMH papers: A personal report. *Family Center Report,* A Publication of the Georgetown Family Center, *15*(3).

Rakow, C. (2013). Analyzing observational data from Bowen's NIMH project: Two months of nursing notes: August 1955 and October 1956. *The Family Systems Forum, 15*(3), 11–15.

ACKNOWLEDGMENTS

No one travels on a journey without help. Many people assisted, at times concurrently, with my efforts.

My greatest debt is to Dr. Murray Bowen, the brilliant original thinker who gave the world a natural systems theory and the groundwork for the next major understanding of life on earth, especially for making sense of human life.

I am so grateful to LeRoy Bowen, who permitted me to review her husband's materials from his time at the National Institute of Mental Health (NIMH) and who generously loaned them to the Georgetown Family Center (GFC). She often brought materials there for review that she thought would be of interest. I returned these materials directly to her.

I thank the Bowen Center for the Study of the Family (formerly GFC) for storing the materials waiting for transfer to the National Library of Medicine (NLM), for encouraging my archival work and for providing a workspace for my efforts.

My appreciation also goes to:

Dr. Michael Kerr, with whom I had that 2011 conversation that gave birth to this book and who also suggested writing a synopsis of the contents of the files, which trained me to think succinctly.

Paul Theerman, Head of Images and Archives in the History of Medicine Division at the NLM, who recognized the extraordinary worth of the Bowen collection and secured permission from Dr. Joanne Bowen, the family representative of the L. Murray Bowen Archives, to transfer the materials to NLM. Dr. Theerman's collaboration with the Bowen Family Trust was crucial to the work of researchers accessing the material when the collection was unprocessed. My work, creating a line-item inventory of the NIMH portion of the L. Murray Bowen collection that serves as a finding aid for other researchers, continued seamlessly when materials transferred to NLM, thanks to Dr. Theerman's oversight.

John Rees, Archivist and Digital Resources Manager in the History of Medicine Division at the NLM, who was professional, gracious and expeditious in accessing needed materials for my references.

Dr. Joanne Bowen, who generously gave me access to the closed collection of papers at NLM and to additional materials then in Williamsburg, VA. Those papers gave insight into her father's life in the army and at the Menninger Institute. Joanne Bowen's personal reflections and her availability to answer my questions advanced my progress and enhanced my personal understanding of the materials. Dr. Joanne Bowen's support has been essential to this effort.

Much appreciation goes to the entire Bowen family. The Bowen heirs' willingness to gift for further study their father's documentation of his odyssey toward science is an inestimable contribution to the future of humankind. It also allowed this researcher to look at materials not yet in the public domain.

The Murray Bowen Archives Project provided lodging, lunch and the opportunity to engage with the archival materials in Williamsburg. This extended the breadth of my research to earlier periods in Dr. Bowen's life and also exposed me to the thinking of other archivists and researchers who were encountering the materials there for the first time.

There are many mentors who were all important in this journey:

Paulina McCullough, at the Western Pennsylvania Family Center, who introduced me to Bowen family systems theory in a way that captured my mind and intrigued me to learn more;

Kathleen Kerr, at the GFC, who suggested that I write to LeRoy Bowen to ask if the paper I was looking for was at the Bowen home—the rest is history;

N. Michel Landaiche, who suggested the perfect title, and whose brilliant mind and understanding of the theory have been integral, even crucial, in guiding, correcting and motivating me to better writing. An unparalleled consultant with an uncanny ability to reveal the ideas in the writing and validate their presence, he advanced ideas and suggested areas for refinement;

Priscilla Friesen, one of the originals to preserve Dr. Bowen's materials, who persevered through years of meetings and lack of funding, human and financial resources while remaining fresh and committed.

Professional writers, colleagues and family all contributed to this book:

Daphne Gray-Grant, publishing coach, whose Get It Done program did exactly that. Having a group of writers to support my entry into the professional world of writing reduced my isolation considerably;

Amie Post, who restored my energy to continue during the days when the possibility I would complete this project seemed nearly nonexistent. Ms. Post's reliability set a standard for readers. She found gaps and unclear passages in my writing but, more importantly, relayed her own experience of learning, giving me a reality check and a road map forward;

Pat Comella, a reader, whose effort was interrupted by life.

Maureen Bayless, a skilled writer and tenacious, resilient, compassionate person, whose gift of time revived my spirit;

Editor Robyn So, gifted, easy to work with and a wizard in her ability to rearrange words. I wish I had met her earlier in the writing process;

Gregg Brown and Joe Brown, for their professional expertise, who polished both prose and references to prepare the manuscript for submission to the publisher;

My son, Paul John, who mentored me in writing citations and in formatting and

My husband, Paul, who never complained about my time on this endeavor. A valued reader, he offered a layperson's perspective. When I was stalled, he was a sounding board, allowing me to talk myself through to new clarity. He tolerated my lack of interest in anything but my work and, many times, fed me lunch when I was lost in the trance-like world of writing. The orientation to facts gained in my effort made it possible to endure his sudden death near the end of writing this book and to continue.

PORTRAIT OF MURRAY BOWEN, 1947

FIGURE 0.1 Portrait of Dr. Murray Bowen in 1947

INTRODUCTION

Bowen's theory foresees a future where the explorations of systems in nature will be at the forefront of understanding this world. When that time comes, Bowen's study of human interaction and the family as a natural system will be a keystone of an all-encompassing understanding of life on planet Earth. Science is on the cusp of finding that "biochemical," epigenetic and "neurophysiological" processes are an outcome of an individual's interaction with an environment that comprised other living systems. Family relations are primary in the human environment. And as a colleague,[1] has asked "Can humans really be understood outside our inter-relatedness with other forms of life (and non-life) that constitute the ecosystems on which we depend for our lives" (M. Landaiche, personal communication, October 10, 2021).

The shift to this paradigm will offer many new strategies for and comprehension of the world. For the human, these will apply broadly from international relations to increasing survival for self and one's family, anywhere humans cluster. This new vista will offer a new world of study based on a history of scientific advance. Life may depend on it. After all, in addition to our behaviors (the outwardly visible manifestations of human movement in the world), the inherent biological processes that are integral to our functioning depend on continued ways of understanding interdependent, living systems.

Organization of the book

This book unfolds as a timeline, considering the world Bowen was born into and the kinds of thinking and resources available to him as he began his professional life and eventually his scientific research. His Army experience and move from

DOI: 10.4324/9781003027287-1

medicine into psychiatry, especially his time at the Menninger Institute in Kansas from 1946 to 1954 (see Figure 0.1, portrait of Bowen, Kansas Historical Society, 1947) and the first year at National Institute of Mental Health (NIMH) are the main emphasis of this book. I devote chapters to his examining of assumptions and forming new concepts. Bowen did not enter psychiatry to create a new theory. Yet, the effort of generating a new scientific theory able to stand the test of time framed Bowen's thinking and behavior. It was at Menninger that the various threads of his interests unified, and he began his trajectory toward a new understanding of the human within the family. After he left Menninger, he established a research project to clinically test and extend his nascent theory at the NIMH.

The focus on ten midlife years, 1946–1955

It was during Bowen's time at Menninger that he formed the foundation on which his theory rests today. In the first year at NIMH, the central building block of the family as a naturally formed unit emerged, allowing Bowen to now explore human behavior under the influence of evolution. Although he continued to revise and focus his theory—to make it more precise—we can, in retrospect, see the seeds of those later modifications in this earlier period. We can see the method of research that he established and refined over his professional life.

Along with a focus on that formative decade, I look at materials from both the preceding and later years, to offer Bowen's reflections on those earlier years and to add context from contemporaneous societal and world events. There are three periods, at least, when Bowen deeply reflects on his place in time. One is 1956–1957, when he had enough confidence to introduce alternative ideas to psychoanalysis by describing the Menninger research for the article "A Psychological Formulation of Schizophrenia," published posthumously (Bowen, 1995). The second is the first half of the 1960s, when Bowen was preparing to publish what became known as his original "six concepts," in 1966. The third was the last decade of Bowen's life (1980–1990), when a retrospective of the theory's development and origins was being prepared for the 1988 publication of *Family Evaluation*, with his colleague Michael Kerr. I use materials from these reviews liberally to enlarge what we can know of the midlife years.

Bowen's work took place in a particular context, at a time in human history of major shifts in our paradigms for understanding life on earth and of key developments in the biological sciences. Bowen was researching, thinking and working in parallel with advances in psychiatry and the broader family therapy movement, some of which I have included to give the larger context for Bowen's early explorations. Events in Bowen's family life pertinent enough for him to mention in his writings and notes are included also. I have referenced presentations Bowen made in those early years and those when he looked back to that seminal decade of researching.

Within the overall narrative, important moments or markers become clear. I think of these as nodal points, "designating a centre of convergence or divergence"

(Oxford English Dictionary, 2003). Bowen's work exemplifies a turning point toward integration of psychiatric thought with the theory of evolution. I hope to show the convergence between Bowen's theory (first formulated or at least imagined at the Menninger Foundation, in Kansas) and the observations and hypothesis-testing he conducted in his first year at NIMH. As he wrote in one of his letters in 1976

> Early in my psychiatric career I engaged a theoretical problem that cannot be answered in one lifetime. In the 1940s I came to believe that emotional illness is far deeper than can be explained by conventional theory. Along with it was a hunch it is related to that part of man he shares with the lower forms of life. These ideas were no more than a distant notion for most of a decade.
>
> *Bowen, 1976a*

Bowen's hunch, along with "educated guess [and] assumption … used to describe a preliminary estimate" (Kerr & Bowen, 1988, 352) had important consequences for how he formed his theoretical model (Bowen, 1976a).

From seeing the emotional system of just the individual to seeing the entire family as an emotional system in its own right was likewise an important shift. In his notes, Bowen described an emotional system as "synonymous with life force— all the way from one-cell animals to the human. Force that causes living things to grow—the force that governs automatic behavior—the force that causes things to grow" (Bowen, n.d.a, 2). This definition encompasses even living things without a nervous system, such as sponges, which have a responsive capacity known as irritability that serves as a guide to survival.

Broadly, an emotional system is a product of nature—molded by evolution and preceding the human. Such a system occurs in every form of life, assisting each to take in information from the environment, coupled with internal information and respond in life-enhancing, life-sustaining ways. A human emotional system involves "automatic" biological processes (e.g., instinct, reproduction, the autonomic nervous system, subjective states). Automatic is the term used to describe responses without volition or conscious control. Much of human life is automatic.

But the emotional system also consists of an intellectual system, e.g., human capacities for thought, reasoning, reflection and some conscious control over automatic reactivity. An ability for some control of automatic responses is unique to the human. Such an arrangement includes the forces governing human relationships, for example, predictable, repeating patterns of interaction, with neurophysiological symptoms signifying a disturbance in the system. The effects of such disturbance or stress are unevenly distributed and non-unidirectional, with individuals varying in vulnerability to absorb the stress for the whole unit.

Bowen considered the human as an instinctual, biological being in a world of other living organisms. In formulating a more precise theory, he drew only from existing scientific theories and findings, excluding anything from fictional literature, the arts, the inanimate world or any other creation derived purely from the human mind. He wanted his theoretical model to be consistent with nature and headed "toward the biological sciences" (Bowen, 1976b). He believed that the human, as a biological being, could be a basis for scientific study. As his research and theorizing continued, he came to see the human family as comprising such a system originating in nature, a natural system.

Bowen showed this determination to move toward science throughout his professional life. One important example occurred in 1972, when selecting guest lecturers for the annual Georgetown Family Center (GFC) Symposium on Family Theory & Family Psychotherapy. There was an intentional shift away from speakers from the family therapy field to scientists and "experts … from peripheral disciplines, who had made a contribution in their own field" (Bowen, n.d.b). Bowen knew these contributions could be of particular interest and relevance to his theory. But I think that contacting leaders in other scientific fields also gave him the opportunity to present his theory to them, a transfer of information that otherwise would likely never occur. By keeping a finger on the pulse of leadership in the scientific community, Bowen fostered a cross-fertilization of ideas and invited the challenges to his thinking that come from having a broader forum.

In describing the planning and implementation of the first year of the NIMH research project, I offer a picture of the first three families who lived on his research ward. I also describe how he gradually began to understand the family as a unit and to work directly with that unit, in dramatic contrast to the psychiatric treatment practices of his day—even of today—that favor working with the one identified, symptomatic patient. I liberally use quotes so that Bowen represents himself on his odyssey. This permits the reader to hear about Bowen's ideas and discoveries directly from him.

The hard part of this effort has been trying to capture the nuances of internal change within Bowen. When he was a resident, I tried to document the challenges he found as he was learning. Far more difficult to record are what he had to unlearn based on newfound evidence. That dichotomy shows up most clearly in his noting his efforts of shifting to a new theoretical base. There are statements that contradict his actions. I've chosen to go with his actions as the emotional understandings, and I considered contradictory statements as old learning not yet updated or survival actions required to navigate the emotional field. As Bowen approached his eighth decade in life, he became much clearer on his historical timetable of developing a new theory.

I close the book with an answer to my motivating question: Where did the ideas for the theory originate?

Beyond this book

There is much about Bowen's long and interesting life that is not found here. This was never meant as a biography. Bowen's research life using archival materials after

1955 is yet to be explored. John Butler researched the origin of family psycho-therapy in Bowen's family study project, but beyond that the field for research is broad and deep (Bowen, 2013). The NIMH project lasted from November 1954 through December 1958, at which time all families were discharged. Bowen left NIMH on July 14, 1959. He began at Georgetown University in September 1959 as Clinical Professor of Psychiatry. By November of that year, he began a formal program on family dynamics and family treatment for psychiatric residents at Georgetown University. He remained on the faculty until his death, in 1990, at the age of 77.

By 1963, more than one source told Bowen that he was close enough to a new theory to put his efforts there (Bowen, 1964). He did. From 1960 to 1966 extensive effort went to defining concepts through clinical practice and research on multi-generational histories of families. He gave up the use of clinical description in his writings, preferring to use theoretical concepts. He expanded the application of the theory to those with less severe emotional problems in his clinical practice and made a number of refinements in the family therapy method consistent with the theory. There were many, many increments in Bowen's thinking before he claimed that he had developed a new theory.

The first publication describing "family systems theory" and its six concepts was "The Use of Family Theory in Clinical Practice," published in *Comprehensive Psychiatry* in October 1966 (Bowen, 1966, 345–374). (This article is also Chapter Nine in *Family Therapy in Clinical Practice*.)

In preparing his chapter for the 1988 book *Family Evaluation*, Bowen reflected on this time:

> In the early years I had hoped the research findings would eventually con-tribute to an extension of psychoanalytic theory and it was important to present the ideas in language the profession could hear. This effort continued into the early 1960s. The last major paper with the descriptive terminology was the chapter in the Nagy-Framo book written in the 1963 period and finally published in 1965. The terminology was retained then because they had specifically asked for a chapter on schizophrenia. In the 1963-1965 period I moved to concepts and for the first time I was willing to call it a theory.
>
> *Bowen, c. 1986, 32*

An NIMH grant for family psychiatry fellowships in 1975 led to the estab-lishment of Georgetown University Family Center, at 4380 MacArthur Boulevard. The move created an opportunity to establish an on-site family clinic and to extend the training program beyond Georgetown University personnel to post-graduate professionals interested in the family. This became known as the GFC.

My research approach

This book represents my efforts to find documentary evidence of Bowen's journey toward science, specifically toward a theory of the human. Very early in my research, I defined my own operating principles when using archival materials. It was important for me to (1) be factual when speaking or writing about this project; (2) maintain the integrity (the meaningfulness) of the composition of the collection; (3) be consistent with the commitment made to this study; (4) understand the collection in enough depth to contribute to a reasoned, organized assembly of the materials for others to use, which organization itself would become part of the broader collection; (5) make known my organizing methods so that others could follow my approach; (6) recognize and acknowledge limitations; (7) seek consultation regarding maintaining archival standards, especially when delineating the emphases that guided my organization of the materials; (8) imagine the collection actively in use by researchers from a variety of disciplines a hundred or more years from now and (9) write regularly on what I was learning. These principles have served me well over the years.

A foray into qualitative research began in 2009, in a course at Union Institute with Michael Quinn Patton, Ph.D., prolific author known for his astute evaluations of programs and organizations, where I looked at two months' worth of nursing notes from Bowen's NIMH project. The first group of notes was from August 1955, when three mother/daughter pairs were part of the project and the concept of the family as a single emotional unit was germinating. The second group of notes was from October 1956, when two mother/daughter pairs remained and two mother/father/impaired offspring families were on the ward, and the effort was to treat the family as a unit (Bowen, 1961, 75). My goal was to be figuratively present on the ward by immersing myself in these selected materials so I could get a hands-on understanding of qualitative research methods and of Bowen's approach in particular.

According to Patton (2002, 4), qualitative analysis seeks to understand the depth and complexity of ordinary human interactions. Toward that end, the researcher can use interviews to create a data set, fieldwork to observe direct interactions among people and written records created by others.

In fact, the nurses' notes contained daily observations of each family member over three shifts, covering 24 hours each day. When I began that study, I wondered what contribution the nurses made to the theory that emerged. Could their notes stand alone as evidence for the subsequent theory? I eventually concluded that those round-the-clock observations made Bowen's new theory possible (Rakow, 2013).

One strength of qualitative analysis is that it can be applied to any data set, and many explorations can use the same data. For example, though I looked at the mother-schizophrenic offspring relationships, someone else examining those same notes might look at the father and impaired offspring; the marital relationship; the mother, father and schizophrenic offspring grouping; impaired offspring and sibling relations; an individual family interacting with staff or family-to-family interactions. The research data can include all of those dimensions.

As I reviewed Bowen's drafts of papers and correspondence from his later years, I gained insight into his earliest operating principles, explorations, hypotheses and theory building. In the pages that follow, I will share some of the gems I found, data that offers a priceless glimpse into Bowen's process of thinking, synthesizing and questioning. I also include excerpts from interviews I conducted with colleagues of Bowen's who worked with him on the NIMH research project.

Note

1 N. Michel Landaiche is an international psychotherapist and group facilitator and is on the faculty of the Western Pennsylvania Family Center. In 2020, he authored *Groups in Transactional Analysis, Object Relations, and Family Systems: Studying Ourselves in Collective Life (Innovations in Transactional Analysis: Theory and Practice)*.

References*

*Unless otherwise specified, the works of Murray Bowen included in this chapter are from the L. Murray Bowen papers, National Library of Medicine, History of Medicine Division, Bethesda, MD. The Accession, Box and Folder information differs for each one and that is included here.

Bowen, M. (n.d.a). [Draft. An overview of the Bowen theory]. (Acc. 2006-003, Box 8, Folder Printed papers in odyssey).

Bowen, M. (n.d.b). [A psychological formulation of schizophrenia]. (Acc. 2006-003, Box 3, Folder A psychological formulation of schizophrenia).

Bowen, M. (1961). Chapter V: Family psychotherapy. In *Family therapy in clinical practice* (pp. 71–90). New York: Jason Aronson, 1978.

Bowen, M. (1964). [Letter to Arthur Rosenthal on March 18, 1964]. (Acc. 2004-043, Box 3, Folder Basic Books Publishing Co., Inc.).

Bowen, M. (1966). Chapter IX: The use of family theory in clinical practice. In *Family therapy in clinical practice* (pp. 147–182). New York: Jason Aronson, 1978.

Bowen, M. (1976a). [Letter to Lewis Thomas on July 24, 1976]. (Acc. 2004-013, Box 3, Folder Symposium 1977).

Bowen, M. (1976b). [Draft, June 1976. Psychiatric contribution to study of the family]. (Acc. 2003-044, box 4, Folder Smithsonian – June 1976).

Bowen, M. (c. 1986). [Draft of epilogue]. (Acc. 2005-055, Box 3, Folder Working papers).

Bowen, M. (1995, Spring/Summer). A psychological formulation of schizophrenia. *Family Systems Journal, 2*(1), 17–47.

Bowen, M. (2013). *The origins of family psychotherapy, the NIMH family study project.* J. Butler (Ed.), New York: Jason Aronson.

Hall, C. M. (1981). *The Bowen theory and its uses.* New York: Jason Aronson.

Kerr, M., & Bowen, M. (1988). *Family evaluation: An approach based on Bowen theory.* New York: W. W. Norton.

Oxford English Dictionary. (2003, December). (3rd ed.). Oxford: Oxford University Press. Retrieved June 19, 2020 from: www.oed.com/view/Entry/127562.

Patton, M. Q. (2002). *Qualitative research & evaluation methods* (3rd ed.). Thousand Oaks: Sage Publications.

Rakow, C. (2013). Analyzing observational data from Bowen's NIMH project: Two months of nursing notes: August 1955 and October 1956. *The Family Systems Forum, 15*(3), 1–2, 11–15.

1

BOWEN'S CHILDHOOD, MEDICAL SCHOOL, INTERNSHIPS

Every person is born into a world of evolving ideas and an accumulation of past investigations and present questions for future examination. We can describe the cultural and economic conditions that Lucius Murray Bowen was born into. Natural systems ideas came readily to Bowen. He credited his early life in a rural area, parents who were in touch with the cycles of life and a feel for nature, and his own personality characteristics as fundamental to that model. And the scientific and psychiatric literature Bowen later read extensively offered him a breadth of knowledge, stimulating challenges, and a depth of influence that are reflected in his writings and which later chapters will explore.

A new century: an integrative process

The first decade of the new century brought the rapid integration of psychoanalysis into the field of medical psychiatry in Europe. In 1902, the Vienna Psychoanalytic Society was created when Freud and four colleagues met each Wednesday in Freud's apartment (Vienna Psychoanalytic Society Records, 1922–1994, n.d.).

The worldwide interest in psychoanalysis to understand human problems and as a viable treatment method prompted two notable occurrences in 1908, one in psychiatry with the beginning of the mental hygiene movement (American Psychiatric Association, 1959) and the other at an international gathering of practitioners, organized by Jung, in Salzburg, for the first International Psychoanalytical Congress (International Psychoanalytic Association, 2019). Jung was a Swiss psychologist who explored consciousness and individuation. Along with another thread of understanding humans from a neo-Darwinian perspective, dating back to 1909, Jakob Johann von Uexküll, who introduced a new scientific paradigm of senses and environment, Umwelt, proposed the idea that "organisms construct their worlds both perceptually and behaviorally" (Deconstructing Niche

DOI: 10.4324/9781003027287-2

Construction: A Conversation between Gordon Burghardt and Kevin Laland, 2017), a notion that I believe fits well with Bowen's eventual concept of differentiation of self. Perception is the filter by which an organism, using all its senses including the cumulative effects of responses to real or imagined threats, assesses its environment; behavior is the action arising from perception. The level of differentiation of self directs perception toward either facts and reality or feelings and visceral reactivity, producing very different behavioral responses.

In 1909, Freud published, "Analysis of a Phobia in a Five-Year-Old Boy," about treating Little Hans using the boy's father as the treatment agent while only once seeing the child. This foreshadowed Bowen's principle that the adult in a family is a better initiator of lasting change in the family, and Bowen's correspondence with a parent without face-to-face meetings early in his research.

That same year, the United States became a leader in psychoanalytic theory (PSA) and practice, expanding the field of child psychiatry when the first child guidance clinic opened in Chicago. In both Europe and the United States, it became an established practice in psychiatry to offer a two-tier approach to problems in a child. While the child received clinical psychotherapy (Zawada, 2015, 25), two other people, a professional and a family member, were introduced into the treatment plan, extending the boundaries of support by offering consultation for parents with the secondary therapist or social worker. While attention went to the healing properties of family relationships, the practice of observing families took place while seeing family members separately.

Amid this two-continent sea change, Lucius Murray Bowen was born on January 13, 1913, in Waverly, Tennessee. He was the oldest of five siblings in a farming family. The world he was born into, so different in its understanding of human behavior from even the ten years preceding his birth, had opened connections between all living things.

Family, medical school, internships 1915–1940

In his developmental years, Bowen's parents were attuned to the life cycles of humans and animals. Farm work was hard and Bowen, as the oldest, had many responsibilities as a farm hand, including chores an hour before walking to school and several hours after it (Bowen, 1979, 2). For a time, three generations lived together as shown in Figure 1.1.

Understanding nature permeated his environment:

> I shall always be grateful for having the parents I did and for the penniless (but successful) situation in which I grew up. I grew up on my mother's peoples farm outside a town of 1,000 .... Mother and her kids raised and preserved all our food except matches, kerosene, sugar and spices. As kids we worked our lives away to help mother prepare the year's food from the fields, garden, orchards, killed farm animals, etc..... Dad was a kind of American Indian who grew up under more primitive conditions but who aspired for the better life.

FIGURE 1.1 Three generations of the Bowen family on the family farm (n.d.)

Source: Reprinted courtesy of the Bowen family.

> He knew animal families from sounds and actions; he KNEW [emphasis in original] botany and what it was about, and he could predict weather from moss on trees, the thickness of animal furs, density of fogs, etc. He knew each and every sound from birds and animals, etc. My brothers and I spent much time with him hunting and fishing for a change in menu on the family table.
>
> *Bowen, 1982, 1–2*

The essence of Bowen came from his position in his family, the orientation of his parents to nature and his own interest in theory and science in those early years (Kerr & Bowen, 1988, 348). Bowen's participation in science and facts goes back to his grade school years from 1919 to 1927 (348).

> I was interested in understanding the nature of the world and in using that knowledge in implementing a better life for the human .... I read everything that came my way, but I was most intrigued by people who had contributed the most in ideas .... [I] grew up on a working farm .... The farm provided immediate access to nature, but the adults in the family worked in town and that provided access to all the inventions of the day. My father was interested in hunting and fishing, but most important was his ONENESS WITH NATURE [emphasis in original] .... I never missed a chance to listen to his explanations of the ebbs and flow of animal life, the weather and our place in the world. He was religious in his own way, but he implied that man fell back on religion to explain things when man's knowledge of facts was

lacking …. I grew up with two major life assets. One was the ability to use my head, and the other was an unusual ability to solve puzzles and use my hands … my ability to tackle a life puzzle, and stay with it … until I had a working solution, was on the positive side …. Most important was an absolute confidence in my ability to think things through and extend knowledge. I believed that man could do anything with his brain if he only applied self to the task. My motivation for that could be called pathological. Through grade school, I looked around for professions or occupation in which knowledge was lacking.

Bowen, n.d.a., 1

His innate abilities sharpened with his childhood reading preferences—inventors, law and medicine, biographies of men such as Walter Reed, bacteriologists, "pure scientists" (Bowen, 1989). These shaping experiences exposed him to the potential of original ideas to extend a body of knowledge.

At the societal level, the early years in Bowen's life were notable for important changes in both the field of psychiatry, within psychoanalysis, and the field of biology. The practice of psychiatry was not well known before the World War I years gave birth to what we now recognize as military psychiatry and the term "shell shock." An understanding emerged of variation in susceptibility to such "wounded minds" (Butterworth, 2018). Psychiatry became important, reaching a comparable status with medicine and surgery. After the war, psychiatry's prominence in the Army receded until World War II prompted its resurgence, in 1941 (Menninger, 2004, 280). And that occurrence involved Bowen.

It was during Bowen's elementary school years that Freud published his concepts of id, ego and super-ego in his book, *The Ego and the Id*, in 1923. "This work contained a final formulation of his structural theory of the mind" (Makavana, 2013).

As Bowen began high school, in 1930, a doctor very important to Bowen's future published his pivotal book and bestseller, *The Human Mind*. Dr. Karl Menninger's book demonstrated that psychiatry was a science (Karl Menninger, 2011) and explained Freudian theory in layperson's terms (Goode, 2003). At the Menninger Clinic, beginning in the 1930s, psychoanalytic ideas were applied with inpatients, a novel use. Karl Menninger had a direct connection with Freudian ideas during Bowen's time at the Menninger Clinic, in the 1940–1950s.

He and Anna Freud openly contemplated psychoanalytic understandings in a lively correspondence between them lasting nearly 40 years (Anna Freud Correspondence, 2007). Anna Freud, a psychologist and a psychoanalyst, brought child psychoanalysis into its own, expanding on her father's ideas with the publication of "The Ego and the Mechanisms of Defence," in 1936, a standard text in psychiatric training today.

Back in Waverly, significant experiences shaped Bowen's future. A well-known story is Bowen's ride in his father's new automobile/ambulance rushing a young woman to a hospital. Bowen drove the ambulance and observed the specialists attempting to help this young woman, who died in the ER despite these efforts. Bowen, at this early age, saw the gap in medicine's knowledge and in the practitioners' application of that knowledge. Writing on the back of the ambulance form, he described the entire scene without interpretation, giving only the facts of the patient's condition, the medical doctor's attendance at the scene, the family's interaction with medical personnel, the ambulance ride and the hospital activities on receiving the patient (Bowen, c. 1939). As an adolescent, he was honing the ability to take in factual information, describe it and challenge conventional understandings.

> The turn toward Medicine occurred early in high school when I watched the staff in a university hospital emergency room fumble with a young woman who had become comatose an hour before. That was a seat of learning in Medicine and the staff did not know what they were doing. The patient died in a few hours later. I came away resolved that knowledge in Medicine was lacking and Medicine would be a good place for my life energy. My decision to go into Medicine never wavered after that.
>
> *Bowen, 1981*

A blend of personal sensitivities contributed to his natural systems orientation: strong self-confidence, an extreme motivation to think problems through to solutions, astute observational abilities and an attraction to strengthening a perceived weak area of knowledge. Bowen recognized the importance of cultivating ideas and traits that precede advancement. How one thinks and approaches questions lays the foundation for what one finds. Such an effort influences one's course. Bowen tilted to the humanitarian side, to contribute to humankind, using awareness of his own strengths and the epic opportunities in what could be known of the human.

Bowen began medical school at the University of Tennessee in September, 1934, with a crucial piece of his approach to understanding and treating human problems already defined:

> A confluence of forces stimulated the long term interest in theory. One was a basic force that had operated even before medical school. It was based on the idea that the specific cause of a medical illness determined the reason for therapy.
>
> *Bowen, 1986*

This was a foundational principle throughout his life: to understand what nature created and how it operates related to human behavior and to plan treatment based on that understanding. The early attraction to science recurred in Bowen's medical school years.

The 1930s saw PSA rise in acceptance, stretching to include the family, and also opposition to it and doubt in the PSA literature among researchers questioning its lack of scientific base. Amid the criticism of Freudian theory that emerged in this decade, Freud announced in 1938 he was standing shoulder to shoulder with other scientists, such as Darwin and Newton (Ritvo, 1990, 198). By 1940, Freud wrote that the stress on the unconscious in psychoanalysis enabled it "to take its place as a natural science like any other" (Freud, 1940, 108).

When Bowen graduated from medical school on September 27, 1937, his interest was surgery, a field moving toward science, and he "was good at it" (Bowen, n.d.b, 2). The specialty fit with his natural interests in problem-solving and his innate ability to work with his hands.

An internship in a small city

The years 1937–1939 were the time of extended learning for Bowen. After graduation, he did a temporary internship in Obstetrics and Urology at Nashville General Hospital, ending in time for him to take the internship exam in December (Bowen, 1941). Foster Kennedy, an innovator, one of the first to use electroconvulsive therapy for psychiatric conditions and to name and delineate shell shock in the World War I (Miller, 2015), was involved in Bowen's Nashville opportunity, yet the archives do not reveal in what way (Bowen, 1941). Kennedy was again influential in Bowen's later Bellevue internship and in his military placement at Ft. Bragg (Bowen, 1988a). Bowen writes, "Among the men who helped shape my life the most, other than my own father was Dr. Foster Kennedy, Chief of Neurology at Bellevue, who had a neurological disease named for him (he was extra special) …" (Bowen, 1988b).

An internship in the country, Cumberland Homesteads

Through the Farm Security Administration, an agency created in 1937 to address rural poverty and one of three agencies administering the Homesteads program, Bowen was in general practice from February to June 1938 at Cumberland Homesteads, four miles south of Crossville, TN (Bowen, n.d.c.). Tennessee was hard hit in the 1930s, not only by the Depression but also by extensive flooding, especially in the Cumberland Plateau region. President Roosevelt, in 1933, had addressed the despair and economic devastation through economic programs intended to give hope, restore stability and offer housing to the loggers, miners, landless farmers and others devastated there. The University of Tennessee and the Tennessee Valley Authority (TVA) gave support and staff to the project (Tollett, 2017a), and by 1934, construction began on Cumberland Homesteads, a residential district of 250 homes made available to those in desperate need once they completed applications and personal interview requirements. The interview included an assessment of the family's moral strength, ethical choices and dedication to being a homesteader, the behavior/conduct of their children and work skills and experiences. "Homesteaders … had to meet rigid requirements of 'high character, ability, honesty and willingness to work and cooperate with the government in this planned community'" (History of Cumberland Homesteads, 2020). Once the evaluation was completed, a trial period of two years took place before the family's final acceptance into the project.

The government advertisement for applicants illustrated that a certain level of maturity was expected in applicants.

> The purpose of Cumberland Homesteads is to give permanently unemployed or displaced families an opportunity to purchase small equipped farms. It is absolutely necessary that each family shall cooperate with the management and other families in building a new community and a new way of living. Anyone not willing to work cooperatively for the good of the entire community should not apply.
>
> *Tollett, 2016*

An example of a family perspective in rural medical practice

The project nurse who did the interviews for the Homesteads program, Amy Cox, is an example of the good nursing that would interest Bowen later in his career. Of the interview experience, she wrote, "Families were visited in their homes. Each family was asked to name three persons as references and families were graded A–B-C-D according to what the case worker observed and the information derived from the references" (Harshman, 1982, 22). An interviewed family member

> recounted how nervous she was when Amy Cox came to do their home visit. The purpose was to see the living conditions in the home and to conduct extensive interviews. They were asked many questions about their backgrounds and questions that would reflect on their character, their work ethic, etc., and detailed questions about their physical condition, such as the ages of each of the children, their height, weight, etc. They were also asked to provide references, and those people were interviewed as well.
>
> *Tollett, 2017b*

Emotional maturity was a criterion in the family interviews. This, along with an emphasis on family, not fathers or job providers, and government responsibly addressing actual need assessed the functioning of the family as a unit.

Bowen boarded at the Cox home during his time working for the project (Bowen, J., personal communication, July 10, 2019). He was the only doctor serving the community's medical needs, including treating pneumonia, which in those pre-antibiotic days was a life-threatening illness. There were mine and construction accidents and farm incidents. The people and their way of life would have been familiar to him. He grew up less than 200 miles from Crossville. Bowen later described himself as having had "[a] rather busy practice of medicine, minor surgery, obstetrics, pediatrics and preventive medicine …" (Bowen, n.d.d.).

Bowen contracted for only half-time general medical practice, so the office hours were brief, 10:30 am–12:00 noon and 3:00–5:00 pm. He also did private work making home visits to families unable to get to the office. Most mountain people did not have cars. He may have seen families who were not part of the

FIGURE 1.2 Cumberland Homesteads office

Source: Reprinted courtesy of the National Library of Medicine, History of Medicine Division, Bethesda, MD.

Cumberland Homesteads project that lived in the rugged hills and valleys of the Cumberland Plateau, including Crossville.

This photo, Figure 1.2, of Bowen outside his office at the Cumberland Homesteads project hung for many years in the Georgetown Family Center (Papero, 2017).

Bowen also maintained a correspondence relationship with Mrs. Cox, even sending her reprints of his papers while at NIMH. In later years, he visited her with his family on trips to Tennessee. Bowen tangibly captured this time in his life when he had crab orchard stone brought from the Cumberland Plateau for the fireplace in his home built in Maryland, near NIMH (Bowen, J., personal communication, July 14, 2019).

Bowen was in the right place at the right time. An attentive, observant person who wanted to understand how human relations work and what builds in success in life, he would learn it in this project. His time in Crossville, TN, and his relationship with Mrs. Cox offered a family perspective on medical practice and the possibility of tracking variation in families that may have influenced future decisions on his path to a new theory.

Experience in a large city, medicine and surgery

The second half of 1938 held moments of uncertainty while bringing important contacts that aided Bowen's career trajectory. His selections show him moving further from Tennessee and rural life. He first tried for an internship at Philadelphia

General, but in New York City serendipitously got three short appointments that he considered first-rate.

The opportunity for a neurology position under Foster Kennedy, Program Director for the Department of Neurology at Bellevue, ended his efforts to get into Philadelphia General. In distinct contrast to the rural environment and poverty of the earlier part of the year, working in the world's largest city, at two different hospitals, with choice options available for his studies gave him a focused experience of inner-city life, exposure to a wide range of the city's population, time under an important mentor, Kennedy, and clarification of where he wanted to go. The Murray Bowen Archives Project notes that working in New York City fulfilled a dream of Bowen's (Murray Bowen in His Time and Place, 2019).

Grasslands internship

Bowen started a two-year internship at Grasslands Hospital, Valhalla, NY, on January 1, 1939. He describes himself on arrival as "a medical idealist who is still in pursuit of the far off goal … the search for the impossible" (Bowen, c. 1969). While there, he was on ambulance service and came to know the surrounding region well (Bowen, 1949, 1). This rotating internship offered equal emphasis on medicine and surgery. As an essential attribute of Bowen's learning method (sort the facts from opinions, feelings, biases and suppositions), he gave due process to medical rotations in all specialties at Grasslands Hospital (Kerr & Bowen, 1988, 348).

> Previous experience enabled many services to be almost that of junior resident. Had excellent two month services in pathology, tuberculosis, psychiatry and outpatient clinics; four month services in medicine and surgery; and shorter services in obstetrics, contagious diseases, pediatrics, ophthalmology, ENT, anesthesiology, urology, gynecology and emergency room. All services closely supervised with active teaching programs and supplemented by participation in the Westchester County Medical Society and the New York Academy of Medicine.
>
> *Bowen, n.d.c.*

The rotating service fit Bowen's interest in learning how the various specialties operated within medicine and knowing how each specialty used a fact-based approach to the art of healing. This anticipated the broader method in his residency at Menninger of searching for facts in diverse disciplines.

There is a strong thread of long-lasting relationships through Bowen's life's journey. In 1970, on the 50th anniversary of Grasslands Hospital, Bowen reflected on his time there:

> I was a crazy medical idealist forced to go to a local medical school because there was no money to go elsewhere. Then I set out to 'find my way'. Bellevue and New York City were disappointing, and I found my way to Grasslands

which, without representing medical excellence, had something that was attractive. I was too 'picky' to accept anything and the way Reid Hefner presented himself and Grasslands had something to do with me accepting an internship there. Once at Grasslands, there were a few medical giants who stood apart from the crowd …. Gilbert Dalldorf stood apart from all the rest as clinical and Scientific [*sic*] giants.

Bowen, 1970

The respect was mutual, as it was Dalldorf who later assisted in Bowen's approval for a fellowship in surgery at Mayo Clinic, in 1940 (Bowen, 1957). Bowen made long-lasting relationships in each of his placements. From early on, others saw something in Bowen that led them to help him advance. This, coupled with his innate ability to maximize experiences, put him in leadership positions.

Surgery was not yet a science but approaching it (348). During the two-month surgery rotation at Grasslands, designing a mechanical heart captivated Bowen's interest, though two months were not enough time to finish it (Bowen, 1977). He said, "This motivation came when the patient died after a complicated chest operation in which the heart could have recovered with a little help with its load" (Bowen, n.d.b., 2).

While Bowen was exploring medicine and surgery, important preludes to his career trajectory were developing. Circumstances in psychiatry were changing. The last half of the 1930s saw an interest in symbiosis in the psychiatric literature. This attention to symbiosis did not represent a shift from an individual orientation to psychiatric problems. Nor did it represent an understanding of a parasitic state of a relationship. Rather, psychiatry understood symbiosis as a developmental lag within the individual (Bowen, 1972). At the end of this decade, William Menninger extended the application of Freudian theory to an institution by setting up

a psychoanalytic hospital in which the total treatment program implemented incorporated approaches to the patients governed by a dynamic understanding of the patients' illnesses.

Menninger, 2004, 278–279

This matched one of Bowen's operating principles. Knowing how something came to be is what guides the treatment choice. Bowen's life experiences through the 1930s paralleled a growing worldwide interest in families and how to understand them, sowing the possibility of Bowen's later career choice into psychiatry.

Life interrupted

The political disturbances in Europe in the 1930s and the start of World War II in September 1939 had a poignant side effect of returning military psychiatry to prominence because of the high number of psychiatric casualties. The day Bowen graduated from medical school, he had accepted an appointment as First Lieutenant

Medical in the Officers' Reserve Corps (Bowen, 1945). Before the Grasslands internship ended, and after the Mayo Clinic accepted Bowen as a surgical fellow, the United States Army interrupted his career plans.

Bowen received notice to report for active Army duty on February 6, 1941, at Ft. Bragg, North Carolina (Bowen, 1941). Foster Kennedy, who had supervised Bowen's internship at Bellevue Hospital, was instrumental in his placement at Ft. Bragg Station Hospital (Bowen, 1988a). There, Bowen's shift to psychiatry became a reality.

References*

*Unless otherwise specified, the works of Murray Bowen included in this chapter are from the Murray Bowen Papers. 1951–2004. Located in: Modern Manuscripts Collection, History of Medicine Division, National Library of Medicine, Bethesda, MD. The Accession, Box and Folder information differs for each one and that is included here.

American Psychiatric Association. (1959). [Impressions from the 115th annual meeting of the American Psychiatric Association, April 27–May 1, 1959]. (Acc. 2006-003, Box 1, Folder Documenta Geigy).

Anna Freud Correspondence. (2007). Kansas Historical Society. Retrieved March 20, 2020, from: www.kansasmemory.org/item/223251.

Bowen, M. (n.d.a). [Draft]. (Acc. 2005-055, Box 3, Folder M Bowen's thoughts on pathway to psychiatry).

Bowen, M. (n.d.b). [Draft: Pathway to psychiatry]. (Acc. 2005, 055, Box 3, Folder M. Bowen's thoughts on pathway to psychiatry).

Bowen, M. (n.d.c). [Letter]. (Acc. 2004, 013, Box 4, Folder Re: Grasslands, correspondence).

Bowen, M. (n.d.d). [Letter]. Copy in possession of Bowen family, Williamsburg, VA.

Bowen, M. (c. 1939). [Ambulance record]. Copy in possession of Bowen family, Williamsburg, VA.

Bowen, M. (1941). [Army document, February 11, 1941]. (Acc. 2007-073, Box 4, Folder Army correspondence).

Bowen, M. (1945). [Form]. (Acc. 200-073, Box 4, Folder Registrar).

Bowen, M. (1949). [Letter on September 28, 1949]. Copy in possession of Bowen family, Williamsburg, VA.

Bowen, M. (1957). [Letter on August 29, 1957 to Gilbert Dalldorf]. (Acc. 2006-003, Box 6, Folder NIMH correspondence A through G).

Bowen, M. (c. 1969). [Draft of letter to colleague from Grasslands]. (Acc. 2007-012, Box 2, Folder Misc. Correspondence).

Bowen, M. (1970). [Letter in response to invitation to attend Grasslands 50th anniversary celebration]. (Acc. 2005-055, Box 4, Folder Grasslands).

Bowen, M. (1972). [Transcript of interview on February 2, 1972]. (Acc. 2007-073, Box 2, Folder APA interview with Mark Johnson, Interview No. 2).

Bowen, M. (1977). [Draft of presentation notes]. (Acc. 2004-013, Box 3, Folder symposium 1977).

Bowen, M. (1979). [Letter to Carl Whitaker on August 14, 1979]. (Acc. 2007-073, Box 1, Folder Name redacted).

Bowen, M. (1981). [Letter, August 7, 1981]. (Acc. 2007-012, Box 2, Folder Bowen re: His hospitalizations in '81, '86, & '90).

Bowen, M. (1982). [Letter on February 16, 1982]. (Acc. 2004-043, Box 4, Folder Dysinger, Robert H.).

Bowen, M. (1986). [Draft]. (Acc. 2006-003, Box 8, Folder Odyssey-in theoretical principle).

Bowen, M. (1988a). [Letter on March 21, 1988]. (Acc. 2004-042, Box 4, Folder (Name redacted).

Bowen, M. (1988b). [Letter on April 26, 1988]. (Acc. 2004-043, Box 5, Folder Belief letters of Dr. Bowen).

Bowen, M. (1989). [Presentation in Ashville, N.C. on September 9, 1989]. (Acc. 2003-044, Box 6, Folder Ashville-Sept 8-10), Paul, Norman.

Butterworth, B. R. (2018, November 8). What WWI taught us. *The Conversation*. Retrieved July 20, 2020, from: https://theconversation.com/what-world-war-i-taught-us-about-ptsd-105613.

Deconstructing Niche Construction: A Conversation between Gordon Burghardt and Kevin Laland. (2017, June 19). Deconstructing niche construction: A conversation between Gordon Burghardt and Kevin Laland. *The Evolution Institute*. Retrieved March 22, 2020, from: https://evolution-institute.org/deconstructing-niche-construction/?source=tvol.

Freud, Anna. (2020, July 15). Freud, Anna (1895–1982). Encyclopedia.com. Retrieved July 20, 2020, from: www.encyclopedia.com/people/medicine/psychology-and-psychiatry-biographies/anna-freud.

Freud, S. 1856–1939. (2020, July 17). Retrieved July 19, 2020 from encyclopedia. com: www.encyclopedia.com/people/medicine/psychologgy-and-psychiatry-biographies/sigmendfreud

Goode, E. (2003, May 31). Famed psychiatric clinic abandons prairie home. *The New York Times*. Retrieved March 22, 2018, from: www.nytimes.com/2003/05/31/us/famed-psychiatric-clinic-abandons-prairie-home.html.

Harshman, A.Y. C. (1982). *I remember.* Byron's graphic arts.

History of Cumberland Homesteads. (2020). History of Cumberland Homesteads. *Cumberland Homesteads*. Retrieved July 21, 2020, from: http://cumberlandhomesteads.org/homesteads-history/.

International Psychoanalytic Association. (2019). Encyclopedia.com. Retrieved July 20, 2020, from: www.encyclopedia.com/psychology/dictionaries-thesauruses-pictures-and-press-releases/international-psychoanalytical-association.

Kerr, M., & Bowen, M. (1988). *Family evaluation: An approach based on Bowen theory.* New York: W.W. Norton.

Makavana, D. S. (Ed.). (2013, October–December). Person of the issue, Sigmund Freud 1856-1939. Academia.edu. Retrieved March 22, 2018, from: www.academia.edu/7281328/THE_INTERNATIONAL_JOURNAL_OF_INDIAN_PSYCHOLOGY_VOLUME_1-ISSUE-1.

Menninger, W. (2004, Fall). Contribution of Dr. William C. Menninger to military psychiatry. *Bulletin of the Menninger Clinic, 68*(4), 277–297. Retrieved from: https://guilfordjournals.com/doi/pdfplus/10.1521/bumc.68.4.277.56641.

Menninger, K. (2011, July). Karl Menninger. Kansaspedia, Kansas Historical Society. Retrieved June 7, 2021, from: www.kshs.org/kansapedia/karl-menninger/17218.

Miller, T. (2015, January 24). The Foster Kennedy House – No. 14 Sutton Square. *Daytonian in Manhattan*. Retrieved July 21, 2020, from: http://daytoninmanhattan.blogspot.com/2015/01/the-foster-kennedy-house-no-14-sutton.html.

Murray Bowen in His Time and Place. (2019). *The Murray Bowen Archives Project*. Retrieved July 20, 2020, from: http://murraybowenarchives.org/murray-New

Papero, D. (2017). Podcast, November 8, 2017. "Dr. Murray Bowen's Growing-Up Years." Interview with Jenny Brown, PhD and Daniel Papero, PhD. From: www.thefsi.com.au/conference-recordings/podcasts/.

Ritvo, L. (1990). *Darwin's influence of Freud: A tale of two sciences*. New Haven: Yale University Press.

Tollett, C. (2016, September 8). Cumberland homesteads—How it all began. *Crossville Chronicle*. Retrieved July 21, 2020, from: www.crossville-chronicle.com/news/lifestyles/cumberland-homesteads-how-it-all-began/article_f4a43250-75fe-11e6-a501-3b5f752aa186.html.

Tollett, C. (2017a, August 17). Cumberland Homesteads: Friends in high places. *Crossville Chronicle*. Retrieved March 2, 2019, from: www.crossville-chronicle.com/news/lifestyles/cumberland-homesteads-friends-in-high-places/article_d7ee146c-8379-11e7-b7d8-ebb22261def0.html.

Tollett, C. (2017b, May 4). Cumberland Homesteads—Our families, Part 2. *Crossville Chronicle*. Retrieved July 21, 2020, from: www.crossville-chronicle.com/news/local_news/cumberland-homesteads-our-families-part/article_b787891e-30ff-11e7-a2a6-4b80b76d1589.html.

Vienna Psychoanalytic Society Records, 1922–1994. (n.d.). (Imaj Associates). Boston Analytic Society and Institute. Retrieved July 20, 2020, from: https://bpsi.org/library/archives/collections/vienna-psychoanalytic-society-records-1922-1994/.

Zawada, S. (2015). An outline of the history and current status of family therapy. In S. C. Box (Ed.), *Psychotherapy with families: An analytic approach* (pp. 25–34). London: Routledge.

2

THE YEARS OF ACTIVE DUTY 1941–1945

Bowen came from the Grasslands internship to the Army (Bowen, 1957) with the expectation his military service would last one year (Bowen, 1941a).

Psychiatry, Registrar and opportunity to develop a broad view

When Bowen reported in February to Ft. Bragg, then the largest living quarters for soldiers in the United States (Tooker, Hartman, & Smith, 2011, 87), he moved straight to the Chief of the Psychiatry department (Murray Bowen in His Time and Place, 2019). Interestingly, no psychiatrist was there.

By July 1941, Bowen became Assistant Chief of Medical Service. Two months later, he was Registrar for three hospitals at Ft. Bragg. By the fall, those hospitals provided for over 250,000 admissions (Bowen, 1942a, 2). With uncertainty and unpredictability in the country and the military, the Army's expectations of its doctors and their terms of duty were in continual flux. Bowen was eager to move into his chosen field of surgery but in September postponed another application to the Mayo Clinic (Bowen, 1941b, 1).

Important relationships in increasing the Army's effectiveness in psychiatry

Norman Brill, Chief of Psychiatry in the 4th Service Command and board-certified in psychiatry, arrived at Ft. Bragg in November 1941. He assumed the position of Chief, allowing Bowen to stay part time on psychiatry under him.

Brill was a close colleague to William Menninger (Coates, 1967), who entered the military after the attack on Pearl Harbor, eventually serving as Chief Consultant to the Army Surgeon General (Menninger, 2004, 279). The relationships between

DOI: 10.4324/9781003027287-3

Brill and Bowen and between Brill and Menninger would be important to Bowen's later decision to go into psychiatry (Bowen, 1943a).

Once the US declared war at the end of 1941, psychiatric services were under increasing pressure to assess and prepare soldiers for combat deployment, then to follow with treatment of the behavioral disorganization and wide-ranging psychological effects of war after deployment. The brutality of war generated many wounds beyond the physical ones, far beyond what institutional psychiatric training provided for.

William Menninger writes, "before actual combat in 1942, there was no definitive plan established to treat neuropsychiatric casualties. It was assumed they could be taken care of in the general hospitals" (Menninger, 2004, 285). Brill and Menninger identified this as a serious gap in planning and set out to remedy this. Brill had written an article in the Army newsletter in November 1941 in which he presciently recognized that the war would bring an absolute need for psychiatrists and gave notice that the Menninger family intended to open a School of Psychiatry (Brill, 1941). The attack on Pearl Harbor occurred one month later, making the need for psychiatrists acute. Having exhausted the available pool of civilian psychiatrists, the Army began its own school for training neuropsychiatrists (Farrell, n. d.). And neuropsychiatry became an official branch of Army Medical Practice in February 1942 (Menninger, 2004, 280). This same year, the Menninger Clinic founded the Topeka Institute for Psychoanalysis (Menninger Clinic, 2020).

An opportunity for research

Even with war preparations ramping up in December 1941, Bowen was confident that his year of duty was ending and sent an application to the Mayo Clinic:

> Made my Mayo application for April 1, 1942 ....The Mayo application went in on my birthday, January 31, 1942 .... I will have the place soon after my Army service is over ... if Mayo will have me, they will get the hardest working man they have had in a long time.
>
> *Bowen, 1942b*

But in the wake of the attack on Pearl Harbor, Congress amended the laws for military service. Bowen's service was extended from the expected one year to the duration of the war, postponing his Mayo fellowship a third time. Ultimately, he would serve five years in the Army, three in the United States and two in England and France.

In Bowen's position as Registrar overseeing three hospitals, he categorized admission rates and hospital turnover. It involved screening soldiers' capacity for Army service and combat duty while assessing categories of disabilities for efficient planning to return to duty. A methodical, disciplined, organized mind and an ability to understand the import of data for planning were essential for that job. The

Army needed to analyze the high turnover of soldiers in their first year of service. Losing soldiers was costly in multiple ways. World War I records did not offer information on this topic, so Bowen was asked in 1942 to track disabilities of soldiers at Ft. Bragg in their first year of service (Bowen, 1943b). The Registrar post was excellent training for evaluating large quantities of data. This major research effort gave Bowen an opportunity to tackle an uncharted area. The Army needed information at once to mount its war effort, organize its operations from induction to service and find potential medical needs that interfered with its goals. Securing and returning soldiers to active service was crucial for the war effort.

What Bowen found was a high correlation between psychosomatic problems and soldier disabilities. This piqued his curiosity for psychosomatic medicine, a holistic approach to body, mind and spirit that anticipated present day bio-psycho-social medicine. Margetts, a Canadian psychiatrist interested in transcultural psychiatry, who searched the history of the word psychosomatic, described it as "… the principle of the organism as a whole, mind and body together, (and) is as old as history itself" (Margetts, 1950, 402). Bowen's work with the Cumberland Homesteads project seasoned him in evaluating strengths of a family for success in that program; the Army research honed his skill in identifying functional limitations and increased his awareness of prevention resources and treatment needs.

Relationships and a crucial decision

In 1943, Bowen worked with Gaine Cannon, "the doctor who put the philosophy of Albert Schweitzer[1] to work in the remote regions of North Carolina" (Blythe, 1964). Cannon and Bowen worked together at the rural medical clinic that was set up by Cannon at Hope Mills, NC, while stationed at Ft. Bragg as a Lieutenant Colonel in the Army Medical Corps (Dorn, 1966). Cannon wanted Bowen to join him after the war (J. Bowen, personal communication, August 24, 2017). Cannon's concept of biochemical homeostasis of the organism, self-regulation that maintains stability, was congruent with Bowen's later interest in understanding adaptive shifts and reciprocal relationships within a family.

The relationship between Cannon and Bowen would have significant implications in Bowen's determinations about his future practice. "Early in my medical career I used to work part time for a general practitioner who was known as the 'Great Doctor' in his community …" (Bowen, n.d.a). Bowen observed that people's belief in this doctor served as the medicine. "I was not about to make new diagnoses or prescribe new treatments" (Bowen, n.d.a). While Bowen treated patients as Cannon did, this partnership gave Bowen insight into his own beliefs, clarifying his principles on his own practice of medicine.

> I can assure you he [Cannon] was one of the most loved and respected men in his community. I resolved this kind of medicine <u>was</u> not for me. He later

had a biography written about him, before his death, which represented him as 'Le Grand Docteur'. For that man to be sued would have been equivalent to trying Jesus Christ for wrong doing.

Bowen, n.d.a

Though Bowen recognized the curative possibilities of being beloved, he also saw its constraint. Gaining a new perspective could be lost in the relationship exchange sustaining such positivity. Bowen's experience as the local doctor for Cumberland Homesteads and his similar experience with Cannon led Bowen to further consider his definition of self as a medical practitioner. His interest was in the mind and body of the human and he wanted to know more facts to further that interest.

The year 1943 was rich in vital experiences that guided Bowen's decision-making. While at Ft. Bragg, because of his family's history as a funeral home owner, Bowen helped the local undertaker, Charles Warren, in Fayetteville, mostly when train wrecks or air crashes occurred (J. Bowen, personal communication. August 24, 2017). Bowen had returned to death practices outside his own family's business, experienced rural medical practice again, and undertaken administrative and research opportunities in his military service. Non-war experiences, psychiatric opportunities in the Army, counterbalanced with combat World War II experiences, provided the warp and woof of Bowen's future.

Planning for the future

Later in life, reflecting on his decision to choose psychiatry over surgery, Bowen expressed satisfaction with his decision, noting that the field of surgery had not seen significant new understandings. Bowen says his "orientation to the human changed" (Kerr & Bowen, 1988, 348) during his military experience. The rise of psychosomatic medicine developed when doctors found no physical cause in at least half of medical problems seen in soldiers. Bowen himself found this in his Army research. Psychiatry had success with those problems (Bowen, 1947, 2). The increase of psychoanalytic practice in the United States during this time was a powerful stimulus to Bowen's musings. The field of psychosomatic medicine emerged along with the acceptance of Freudian theory.

From June 1943 to January 1944, Bowen served as the Ward Officer on the psychiatric service under Major Brill as Chief of the Service (Bowen, 1943c). He was deeply conflicted on what direction to continue his career. In a June 1943 letter to the family, he writes,

one of the toughest decisions I have ever had before me is here right now and I must make up my mind soon .... So far I have been considered a psychiatrist and so far they have planned to make a psychiatrist of me but I can hardly get surgery out of my system. It all boils down to the fact that if I do go to Mayo's in Surgery after the war is over, it will take 5 years before I can go to

> practice. If I switch to Medicine …, I can get to practice a lot sooner and still achieve some degree of prominence in the specialty.
>
> *Bowen, 1943d, 2*

Bowen's concerns show the reality of his Army service, his career ambitions and his practical nature. In July 1943, Bowen writes to his family that he likes psychiatry and his expectation is "I will be able to hold my own" in that specialty (Bowen, 1943e).

Psychiatric problems in soldiers were overwhelming during the war. In Bowen's experience, both psychiatry and surgery lacked confidence in their practices. Yet psychiatry claimed success in the methods of Freud and in a new scientific approach for the post–war era.

> In the army, I was one of the men who became impressed with emotional problems and the comparative helplessness of organically orientated medicine to deal with them. My association with Dr. Norman Brill at Fort Bragg undoubtedly helped kindle my interest, but by the time the war ended, I was sure that my interests were in psychiatry and set out to find a place for training.
>
> By 1943, after I had met so many 'normal' people in psychiatry, and the field became more palatable to me, I was sure I would go into psychiatry after the war.
>
> *Bowen, n.d.b, 6*

The year 1943 was notable in psychiatry for the transition from "shell shock" (Roberts, n.d.) to battle fatigue or combat "exhaustion" (Crocq, & Crocq, 2000) in disabled soldiers. These terms were the precursor to post-traumatic stress disorder (PTSD), which was acknowledged as a medical disorder in 1980 and included as a diagnosis in the American Psychiatric Association's Diagnostic and Statistical Manual of Mental Disorders, the DSM-III (Friedman, 2019).

Bowen decided psychiatry held an appeal over surgery. Brill recognized Bowen's skill inviting him, in November 1943, if he might consider going to Charlotte, NC, to open a psychiatric facility when the war ended (Bowen, 1943f), Instead, Bowen was considering further training. "Brill knows Col. Menninger well and says he thinks he can get me a fairly good place if I want to change. I told him to see what he can find" (Bowen, 1943g). This comment corroborates the assertion that Bowen's change of plans from surgery to psychiatry occurred before he went to the front lines.

Overseas war time duty

During the accumulating war pressures, in early December 1943, Bowen let his family know he intended to marry and did so four weeks later (Bowen, 1943g). The Army alerted him to overseas duty this same month (Bowen, 1943b).

As Bowen headed overseas to England in February 1944, Brill moved from Ft. Bragg to Washington, DC, as Chief of Psychiatry at the Office of the Army Surgeon

General, under William Menninger. The Brill–Menninger association was a prime acknowledgment of post-war treatment needs in psychiatry.

Bowen had favored continuing psychiatric work once overseas. The Army made him Chief of the Venereal Disease Section. This gave him a unique experience that he found interesting and was illuminative in his later work with alcoholics (Bowen, n.d.c, 2). As part of his Army duties in this position, Bowen conducted seminars on Intensive Arsenotherapy in Europe. Arsenotherapy is a treatment of syphilis using arsenic. This experience caused him to pay close attention to the body as a physical system.

> I had a memorable experience using poisonous drugs ... in England during WW II. I was Chief of VD and syphilis was rampant ... the only effective treatment was the arsenical drugs, used in small doses over a long period of time, because arsenic is sort of poisonous. Result was that syphilis became latent, but lifelong. The Chief of Syphilology ... devised a plan to put 3 yrs. of therapy into 3 wks ... called "Intensive Arsenotherapy". Patients put into the hospital, with liver and kidney checks every day, and a dose of arsenicals every day that equaled an average monthly dose. Most soldiers tolerated it very well, and the percentage of complete cures with new cases went into the 80 to 90% range. My job was a tight wire act .... Experience provided courage .... The Intensive Arsenotherapy was a major medical breakthru that was never really recognized. Penicillin came out in the middle of 1944 and everything shifted to that. Early penicillin supplies went first to the Armed services, first to serious medical problems and then to VD in the Armed Services ... the world never really heard about the innovative method developed .... That experience taught me that the body can handle far more toxic substances than we ever believed was possible. Not many people were involved in that arsenotherapy program – maybe a dozen of us, each based in a separate General Hosp. classified as a "VD Center". That was 37 years ago. Maybe some of that knowledge has filtered down to the present generation of "chemo" therapists who are many times more sophisticated than us fumbling Army doctors in 1943-44.
>
> *Bowen, 1981*

This experience foreshadowed Bowen's own work in his regression program with alcoholics and paraldehyde.

By August, Bowen was in France, deep in the war zone. The many wounded soldiers in the Battle of the Bulge (December 1944–January 1945), the last major German offensive in Western Europe, came to Bowen's Division. In a poignant segment of his 1977 Christmas letter, Bowen wrote of Christmas in 1944, reflecting what many soldiers experienced:

> The Battle of the Bulge in early December 1944 was near us. I was the medical officer in charge of R & D (receiving and disposition). Xmas Day

> 1944 was memorable. Hospital trains were coming from the front (350 litter patients per train) once or twice a day and we were loading about 1 train load a day for the hospitals in Paris, or for return to the U.S. I worked all day Dec 24th, all that night, and all day Xmas Day. About 11 p.m. Dec 25th, after I checked in the last trainload of new patients, there was time to go to the Officer's Club for a drink. We had a 'special' dinner in the Mess Hall that day. That Xmas 1944 was the most 'No Christmas' in my life.
>
> *Bowen, 1977, 2*

Bowen returned to the states in September 1945, the same month that representatives of 26 countries established the United Nations with an aim to prevent future wars. In October, Bowen was discharged from Army service, with a leave expiring in January 1946. He had served from February 1941 and had advanced in rank from 1st Lieutenant to Major. The Army recommended him for two Bronze Star Medals for his 95th General Hospital service (Army Document, 1945). At discharge, his superior described Bowen as "a gentleman of the highest caliber, an officer highly capable and competent with a splendid personality and sense of humor. We shall miss him and cherish with pride his great contribution to the 220th General Hospital" (Army Document, 1945). From top to bottom of the hierarchy, Bowen was highly respected.

War time experience changed Bowen's beliefs, feelings and disposition to the human (Kerr & Bowen, 1988, 348), and unexpected circumstances shaped his trajectory. Congress had altered his plans to go to the Mayo Clinic; holding significant positions in psychiatric services, he had learned of human functional capacities in the Army research he conducted and in his position using Arsenotherapy; he had seen the devastation of war and the magnitude of human resilience a thousand miles from home. His unwavering odyssey toward a factual theory began on discharge from the Army but layered on his prior life experiences. His experience with psychiatric casualties and the lack of surety in psychiatry compared to general medical or surgical departments profoundly influenced him.

Moving forward

After Bowen met his wife at Camp Shelby, MS., the two of them headed east in search of a residency program in psychiatry (Bowen, 1977, 2). He was considering the Psychiatric Institute at Columbia, Payne Whitney and Bellevue in New York or programs in Tennessee and Washington, DC. They traveled to see their respective families, then to New York City and to Washington, DC, arriving in November. It was there he heard of the VA-Menninger collaboration. Though wary of VA involvement, the pay was more reasonable than other places. He now had a wife to consider. Bowen visited Brill, who highly recommended the program at Menninger and volunteered to send a letter of recommendation (Bowen, 1977). The Menninger Clinic, with its Topeka Clinic of Psychoanalysis as part of its services, was well-known for Freudian thinking. In December 1945, Bowen explored the opportunity at Topeka, staying two weeks "until I had a promise of acceptance, and we had a room to stay in …" (Bowen, 1977, 2). Bowen was the first to take the psychological exam for residency applications at the new Menninger School of Psychiatry. He met with the screening committee, but Karl Menninger, William's brother, was ill and unavailable. After several days of waiting, Bowen refused to leave without an answer, finally being sent to Karl Menninger's bedside. Their discussion cycled through war experiences and to both Brill and William Menninger. Once Menninger heard that Brill had sent Bowen to him, he was in (Bowen, 1984).

Note

1 "Albert Schweitzer" was a theologian, writer and medical doctor who believed that a "reverence for life" would "naturally lead humans to live a life of service to others." www. history.com/this-day-in-history/albert-schweitzer-born. Cannon built the only hospital Albert Schweitzer would allow to be named for him. It was built from stones offered by Cannon's patients in lieu of payment over many years (Dorn, 1966).

References*

*Unless otherwise specified, the works of Murray Bowen included in this chapter are from the Murray Bowen Papers. 1951–2004. Located in: Modern Manuscripts Collection, History of Medicine Division, National Library of Medicine, Bethesda, MD. The Accession, Box and Folder information differs for each one and that is included here.

Army Document. (1945). [Headquarters 220th General Hospital September 22, 1945]. L. Murray Bowen papers, History of Medicine Division, National Library of Medicine, (Acc. 2007-073, Box 4, Folder Army), Bethesda, MD.

Blythe, L. (1964). *Mountain Doctor: The doctor who put the philosophy of Albert Schweitzer to work in the remote regions of North Carolina*. New York: William Morrow.

Bowen, M. (n.d.a). [Presentation, Some psychological aspects of malpractice suits]. (Acc. 2007-012, Box 3, Folder Science and therapy, Third Thursday Lecture).

Bowen, M. (n.d.b). (Acc. 2007-073, Box 4, Army Correspondence).

Bowen, M. (n.d.c). [Letter to General]. Copy in possession of the Bowen family, Williamsburg, VA.

Bowen, M. (1941a). [Army document]. (Acc. 2007-043, Box 4, Folder Registrar Medicine), (Acc. 2007-073, Box 4, Folder Army Correspondence).

Bowen, M. (1941b). [Letter to Donald C. Balfour on September 28, 1941]. (Acc. 2007-073, Box 4, Army Correspondence).

Bowen, M. (1942a). [Paper Read by Captain L. M. Bowen, MD Station Hospital, Fort Bragg, North Carolina before conference of Chiefs of Services of Fourth Service Command at Atlanta, Georgia on November 21, 1942]. Copy in possession of the Bowen family, Williamsburg, VA.

Bowen, M. (1942b). [Letter on February 9, 1942] Copy in possession of the Bowen family, Williamsburg, VA.

Bowen, M. (1943a). [Letter on August 7, 1943]. Copy in possession of the Bowen family, Williamsburg, VA.

Bowen, M. (1943b). [Letter on December 6, 1943]. Copy in possession of the Bowen family, Williamsburg, VA.

Bowen, M. (1943c). [Letter on June 15, 1943]. Copy in possession of the Bowen family, Williamsburg, VA.

Bowen, M. (1943d). [Letter on June 21, 1943]. Copy in possession of the Bowen family, Williamsburg, VA.

Bowen, M. (1943e). [Letter on July 23, 1942]. Copy in possession of the Bowen family, Williamsburg, VA.

Bowen, M. (1943f). [Letter on November 26, 1943]. Copy in possession of the Bowen family, Williamsburg, VA.

Bowen, M. (1943g). [Letter on November 14, 1943]. Copy in possession of the Bowen family, Williamsburg, VA.

Bowen, M. (1947). [Letter on February 17, 1947]. Copy in possession of the Bowen family, Williamsburg, VA.

Bowen, M. (1957). [Letter to Gilbert J. Dalldorf on August 29, 1957]. (Acc. 2006-003, Box 6, Folder NIMH Correspondence A Through G.).

Bowen, M. (1977). [Letter on December 25, 1977]. Copy in possession of the Bowen family, Williamsburg, VA.

Bowen, M. (1981). [Letter to a psychologist on January 9, 1981]Clinton Phillips. (Acc. 2003-044, Box 5, C.F.S.C. – Dec. 11-14, 1980).

Bowen, M. (1984). [Letter]. Copy in possession of the Bowen family, Williamsburg, VA.

Brill, N. (1941). [Letter]. Copy in possession of the Bowen family, Williamsburg, VA.

By-Laws and Constitution. (n.d.). By-laws and constitution. *Society of Biological Psychiatry*. Retrieved August 27, 2019, from https://sobp.org/membership/by-laws-and-const itution/.

Coates, J. M. (1967). Medical Department United States Army in WWII. Washington, DC, National Library of Medicine. Retrieved March 10, 2020, from https://collections.nlm. nih.gov/ocr/nlm:nlmuid-0211560X1-mvpart.

Crocq, M. M. A. & Crocq, L (2000, March). *From shell shock and war neurosis to posttraumatic stress disorder: A history of psychotraumatology. Dialogues in clinical neuroscience*, 2(1), 47–55. https://doi.org/10.31887/DCNS.2000.2.1/macrocq. US National Library of Medicine. NCBI National Center for Biotechnology Information. Retrieved August 27, 2019, from www.ncbi.nlm.nih.gov/pmc/articles/PMC3181586/

Dorn, H. B. (1966). Mountain Doctor. In *Congressional record appendix* (vol. 149, no. 168). Washington, DC: United States Congress, 2010.

Farrell, M. M. (n.d.). *Chapter III, Office of medical personnel* (p. 43). US Army Medical Department, Office of Medical History, Washington, DC. Retrieved March 2, 2020, from https://achh.army.mil/history/book-wwii-neuropsychiatryinwwiivoli-chapter3

Friedman, M. P. (2019, October 14). *PTSD history and overview*. US Department of Veteran's Affairs. Retrieved September 7, 2020, from www.ptsd.va.gov/professional/treat/ essentials/history_ptsd.asp.

Kerr, M. & Bowen, M. (1988). Epilogue, An Odyssey toward science. In *Family evaluation: An approach based on Bowen theory*. New York: W. W. Norton.

Margetts, E. L. (1950). The early history of the word "psychosomatic. *Canadian Medical Association Journal, 63*(4), 402–404. PMID: 14778085; PMCID: PMC1821745.

Menninger, W. W. (2004, Fall). Contribution of Dr. William C. Menninger to military psychiatry. *Bulletin of the Menninger Clinic, 68*(4), 277; ProQuest Central.

Menninger Clinic. (2020, September 7). Retrieved September 7, 2020, from Encyclopedia. com: www.encyclopedia.com/psychology/dictionaries-thesauruses-pictures-and-press-releases/menninger-clinic.

Murray Bowen in His Time and Place. (2019). Retrieved September 7, 2020, from The Murray Bowen Archives Project: http://murraybowenarchives.org/murray-bowen-in-his-time-and-place/.

Roberts, A. (n.d.). *Mental health history timeline, 1943*. London: Middlesex University. Retrieved August 27, 2019, from http://studymore.org.uk/mhhtim.htm.

Tooker, M. W., Hartman, E., & Smith, A. (2011). *Fort Bragg old post historic district landscape report*, US Army Garrison, Fort Bragg, Cultural Resources Management Program, Building 3-1333, Directorate of Public Works Fort Bragg, N.C. 28310. Ft. Bragg, NC.

3

BOWEN'S RESIDENCY SCHOOL OF PSYCHIATRY AT MENNINGER FOUNDATION, 1946–1947

Bowen was just the type of applicant the new School of Psychiatry wanted, and he met the Menninger's admission standards. There was excitement about the new direction in his life, although bad weather and a need for tire chains kept that in check when he and his wife set out for Topeka in December 1945. "The trip was terrible and slow with much snow" (Bowen, 1977b, 3).

Menninger Complex

Before the war's end, Karl Menninger negotiated with the federal government to allow the Winter General Hospital in Topeka, used by the Army during the war as a psychiatric hospital for soldiers, to be converted to a Veteran's Administration Hospital (Kansas Historical Society, Menninger Clinic, 2004) and become part of the training program at the proposed School of Psychiatry. The Menninger School of Psychiatry, founded by Karl and Will Menninger, and the Winter Veterans Hospital Acute Psychiatric Services opened in January 1946.

Besides the VA Hospital, the Menninger Foundation had two hospital buildings known as the East and West Lodges that served as a sanitarium for long-term care. These were locked units. Supporting the Menninger's respect for relationships, the Foundation had an early model of half-way houses. A converted farmhouse, renamed The White Cottage, was an open setting offered as a continuing care residence, much like today's step-down unit might serve to enable reintegration into society for those no longer needing intensive in-patient care but not yet ready for independent living (Schlesinger, 2007, 233). As a next step, the Menninger Foundation negotiated with families in the local Topeka community to offer lodging for those patients well on their way to recovery. This gave proximity to supportive relationships and easy access to out-patient treatment at the Clinic. A social

DOI: 10.4324/9781003027287-4

worker held monthly meetings with the families and provided supervision of the people living with them (Schlesinger, 2007, 233).

The great need for psychiatrists after the war came with the recognition of battle exhaustion. The new school would support informal research and an adapted Freudian approach to treating the "whole" person in the context of long-term hospitalizations (Smarsh, 2010, 93). Starting with 100 psychiatric residents and 10 psychologists, the school would graduate clinical staff trained in this approach. The Menninger brothers had long prepared for this. Both Menninger brothers, in 1934, sought the advice of European experts about creating a much larger facility. Dr. Karl met with Sigmund Freud about this. Freud did not encourage treatment in a hospital environment. Dr. William met with Ernst Simmel, German neurologist and psychoanalyst, who had pioneered the idea of a hospital using psychoanalytic treatment within a simulated family environment. The meeting with Ernst Simmel, influenced the methods chosen for the milieu. Simmel used staff to function as family surrogates (Mackie, 2006, 6–7; Mackie, 2016, 107; Schlesinger, 2007, 230). Mackie, an Australian psychoanalyst, describes this approach.

> Clinically the sanatorium provided a total therapeutic milieu that comprised a staff of analyst physicians, a matron, analytically trained medical assistants …, nurses, house personnel and a manager …. The clinic developed as a sort of extension of the analyst function and as the archetype of the family in general. The matron and attendants represented the mother, the physician/ analyst the father and the other patients and sometimes nurses standing in for siblings. Every morning Simmel would meet with the staff to discuss the patients and give directions for the day. He paid close attention to the feedback from hospital staff who he referred to as an 'extra sense organ' for the analyst with regard to conflicts that were projected onto the staff in the transference.
>
> *Mackie, 2006, 6–7*

Transference is a Freudian term. When the patient interacted with the analyst using the same response from parent–child interactions, this was known as "trans-ference." Resolving a transference was the bread and butter of psychoanalytic treatment. The Menninger approach reframed clinical treatment from the indi-vidual and psychotherapist to creating a healthier living environment that would benefit the individual in interaction with the staff (Mackie, 2016). This meant ALL staff, from physician to aides, dietary personnel, clerical staff and house cleaning were part of the therapeutic team. Herbert Schlesinger, a psychologist starting at the Menninger School of Psychiatry when it opened in 1946, wrote about the expectations for staff:

> The nurses, psychiatric aides, occupational therapists and recreational therapists, the nutritional staff members, and members of the clerical staff,

effectively were on the therapeutic team. All of them knew the patients well and were expected to understand how to apply the program. The therapeutic intent was not to promote transference .... The hospital milieu was [to assist] the patient to develop a pattern of healthful living, and to reverse and correct the distortions of personality that the contemporary theory of psychoanalysis held responsible for the illness.

Schlesinger, 2007, 230

The therapeutic milieu was adapted by the Menninger brothers from the model for psychoanalytic treatment based on Freud's study of two-person relationships. Mary Bourne, one of the earliest non-physician students of Bowen theory, described the analytic method thus:

The goal of the analyst is to build on that 'transference' relationship, intensify it and use it in psychotherapy, then gradually move away from it so that the patient becomes more of a self, and the analyst also separates out a self in relationship to the patient.

Bourne, 1987

Bourne captured the reciprocity in the transference. It is the interaction between the two people, one bringing the remnants from the original transference (a third person) and one aware of the process taking place and guiding that process to separation. At the Menninger clinic, all staff were to be aware of that process and provide interactions intended to foster that process freeing the person to try new interactions.

Bowen was a good emotional fit with the Menninger brothers' approach to care of those people having extreme struggles in life. The Menninger Clinic and Foundation were a family affair. Their orientation to milieu, family and treatment fostered Bowen's innate systems thinking.

Bowen was the first resident to arrive at the program, the day after it opened on January 2, 1946 (Bowen, 1947a). There were 108 physicians and clinical psychologists in the first class, with hopes of a cross-fertilization of research and clinical practice with psychoanalytic teaching (Schlesinger, 2000, 52).

Residents accepted in the School of Psychiatry were not called "residents" on entry. They were Junior Medical Officers in the VA. For his first six months, Bowen was Acting Chief of the newly opened Winter Veterans Hospital Acute Psychiatric Services (Bowen, 1946a), then the largest psychiatric training center in the world. Neither it nor the Menninger in-patient sanitarium offered formal psychotherapy as a given. Patients had to request it. Beginning as a Fellow, Bowen's position included daily contact with patients that offered them a brief psychotherapy experience. He had responsibility for evaluating a patient's status and had transactions with nursing, occupational therapy (OT) and recreational therapy (RT) staff. One expectation of the position was "a continuing evaluation of the patient's progress" (Schlesinger, 2007, 234). This meant defining exactly what progress is.

After the six-month orientation to policy and practices, in June 1946, Bowen and a few others in the class were placed in service at the Menninger Clinic. They became official residents in September 1946, though the new title came with a pay decrease (Bowen, 1946b). In December 1946, Bowen was invited to be on staff at the clinic as a Jr. Psychiatrist, becoming "the 1st VA resident promoted to a staff position at The Clinic" (Bowen, 1984). Bowen remained a full-time staff member, serving successively as Junior, Intermediate and Senior Psychiatrist and Section Chief at Menninger Clinic, and as Clinical Instructor in the Menninger School of Psychiatry, until he left for National Institute of Mental Health (NIMH) in June 1954.

Life as a resident

Once full-time in the residency program, Bowen's motivation was to learn much and to learn fast, a young man eager to get into his career. World War II had interrupted and slowed his life plans.

On Saturday mornings, the residents in psychiatry and clinical psychology met in Karl Menninger's Colloquium (Schlesinger, 2007, 235) for courses in clinical psychology. The Colloquium benefitted from the rich experiences of those who had trained in Europe. Psychoanalysts, either on migration to the United States or through solicitation by the Menninger brothers, found positions here. Ellen Simon, who later became Bowen's analyst, was one. Rudolph Ekstein, a psychoanalyst trained in Vienna who fled in the wake of Hitler's invasion of his country became the head of the education department. He later supervised Bowen as he completed his clinical control cases for credentialing as an analyst.

Bowen used his time as a resident to develop researchable questions by learning as much Freudian theory and psychoanalysis (PSA) as possible. He appreciated the opportunity for theoretical as well as practical training; had he chosen surgery, as he once planned, he realized it would not have had this potential for diving into the theory underlying the practice.

Bowen's first questions concerned the basic definition of certain terms. Clarity of meaning engaged his innate interest in science and problem-solving. He questioned instructors and proposed ideas for language that would work better. By gravitating to problem definitions, though, Bowen was on the fringe of Freudian theory. The responses of instructors to Bowen ranged from parody to respectful appreciation.

Bowen's studies

When Bowen applied to Menninger, his understanding was that Freudian theory was scientific. On arrival at Menninger, a report caught Bowen's attention that questioned whether Freudian theory was as scientific as reported (Kerr & Bowen, 1988, 345). "This one sort of was a surprise. I had expected human behavior, the human being, to be as much of a science in psychiatry as in medicine or anything else" (Bowen, 1987, 7). Bowen's interest in ideas—that how one thinks guides what one does—had met with a need that proved both a challenge and an opportunity.

Bowen referred to these years as the "first phase" in his odyssey toward a science of human behavior (Kerr & Bowen, 1988, 356). Initially excited with the understandings of human behavior found in psychoanalytic explanations, within a year he was aware of discrepancies in the theory. The criticism that psychoanalytic theory was not a science prompted Bowen to investigate what it would take to make it so (Bowen, 1977).

Writing on what drives science, Stuart Firestein, chair of the Department of Biological Sciences of Columbia University, captured the essence of Bowen's efforts a half-century later: "One good question can give rise to several layers of answers, can inspire decades long searches for solutions, can generate whole new fields of inquiry, and can prompt changes in entrenched thinking" (Firestein, 2012, 11). Bowen's questions about psychoanalytic theory and accepted knowledge led him directly to the sciences, evolution and the family. In putting psychoanalytical principles to the test of fact, he could separate out those principles from literature or the arts. Thus began a lifelong quest to advance a theory of the human based on the science of nature. In 1946, however, he thought, "It won't take long … to find out where Freud missed science" (Bowen, 1990, time stamp 4:23).

Data on families

As part of their milieu, Menninger had an extensive preadmission evaluation for patients:

> On admission, a patient and accompanying family members engaged in several weeks' long evaluations of the patient, his family, and his social, work and educational settings. A team of psychiatrist, clinical psychologist and psychiatric social worker conducted this diagnostic study. Although the nature of their interactions with the patient might overlap considerably, each member of the team had clearly defined functions.
>
> *Schlesinger, 2007, 233–234*

Bowen found that even though extensive data was gathered on families' intake assessments, his observations of the family did not match what he was being taught in his classes at the School of Psychiatry. This is an example of Bowen's efforts to move toward science. Addressing this mismatch by adding coherence was a place to start.

As early as 1946, he experimented with the analytic tenet that contact with family contaminated the inviolability of therapist/patient boundaries. He sent regular mailings to families, providing them with information on how the course of treatment with the hospitalized family member was going. He maintained contact on discharge. This opened interactions between Bowen, the family and the patient. The patient, in a clinical session, spoke about how the family received the letter.

This allowed Bowen to discover for himself how family interaction reports differed from family member to him and from patient to him—a very useful method of learning the reality of the nuclear family interface. Bowen used letter writing to keep contact with people after discharge:

> I began an informal clinical experiment at the Menninger Clinic to encourage extensions of psychotherapy relationships by mail. Many of the people who sought help at the clinic were from distant places. As a Section Chief, I had clinical contact with a large number of people. When I left Topeka in 1954, I continued the relationship with those who wished to continue .... The main goal of the study was an effort to make meaningful psychiatric help available to more and more people with less time ... a psychiatrist can work with no more than a few hundred people in a lifetime .... I believe that, if I could make just one contribution to psychiatric knowledge, I would have contributed more to the overall cause of mankind than would ever be possible in a life time of work with individuals.
>
> *Bowen, 1959*

Long before computers enabled virtual consultation, Bowen had a method for helping people around the country. A clinical extension of using letters for family contact, out-patient supportive psychotherapy by mail exchange developed and continued in later years with in-person sessions occurring once or twice a year. Bowen identified the difficulties of personal contact by mail.

> The main advantage of the mail relationship is the freedom from schedule and travel problems. From a practical standpoint, it was possible to maintain an active, effective relationship by mail. There were several disadvantages. The correspondence relationship requires quite a bit more skill than the 'alive' face-to-face relationship. There is not the opportunity to ask questions about vague, overlapping, and oblique communications. It is easier to 'guess at' what the other meant than to write a letter to ask for clarification. Many times, I have put more time and energy into trying to piece together the information in a letter than would have gone into a psychotherapy hour. Also, on the disadvantage side is an unavoidable skewing of the emotional process. In the 'alive' situation, people have ways of knowing the emotional process in the other, even without an exchange of words .... Moods and life situations can change during the time required for an exchange of letters .... I have had correspondence relationships terminate abruptly over the misunderstanding of a vague communication. The face-to-face situation can tolerate much more misunderstanding. The one biggest advantage is the amount of time the therapist can put into this effort if the goal is the maintenance of a psychotherapy relationship, the response is quite a bit more complicated than the mere answering of a letter.
>
> *Bowen, 1959*

Some recipients of Bowen's letters maintained a correspondence with him until his death in 1990. This approach, offering a professional resource to those wanting psychiatric help well beyond traditional office-only relationships, has made processing Bowen's archives to current Public Health Information (PHI) standards a challenge. Every piece of correspondence has the potential to be considered a clinical session.

Course work, questions and contradictions

Psychoanalytic theory was based on work with neurotic-level problems. Bowen asked, was the theory too difficult to apply to more serious diagnoses or was the theory actually not even applicable? Bowen chose an in vivo path to discover answers. He had read what others thought and conceptualized and prodded teachers for answers on discrepant ideas in the theory. The response to him was often that if he understood Freudian theory, such questions would not exist (Bowen, 1973, 6). His immersion in Freudian theory showed where gaps in the theory existed and a direction toward more substance. The lack of examples of accepted theoretical tenets, such as maternal deprivation and the clinical provision of a corrective emotional experience, each of which had wide variance in outcomes including no symptoms, left space to think for himself and space for clinical exploration. Other causes of symptom eruption, such as traumatic experiences, often happened to people never hospitalized. Those inconsistencies turned him toward a lifelong principle of creating circumstances where he could investigate for himself, so answers came from the definitive source, the patients (Bowen, 1973, 6). As they were observed, the patients themselves could teach him and lead him to new questions:

> 1940s–[I was] interested in schizo[phrenia] – Institution acted as if it knew precisely what to do to modify schiz[ophrenia] – One p[atien]t a great teacher – Had staff help him get dressed, etc., before going on grounds–[He] Would get outside the door and promptly mess up hair & dress–take off 1 sock, take laces out of shoes–etc. [I] Tried 'reversal' or 'strategic' therapy– Instead of fixing him up, the staff messed him up – Outside the door, he would 'fix' himself up. How did one figure this?
>
> *Bowen, 1985*

Instead of blaming the patient's personality flaws, Bowen considered something going on in the relationship with staff. What would be more effective? Trying to generate insight within a patient did not bring solid change. A patient's own actions arising out of thoughtful self-assessment could make a difference, and staff moving toward the patient to "help" him fostered a reversal toward the staff. It was an inductive approach to form his own questions in his work with patients. Some teachers viewed him as unwilling or unable to grasp the theory when he questioned psychoanalytic readings inadequacies to answer his questions. There were times when he could be humorous with his instructors. On a handwritten test, when asked to give definitions of six mechanisms of defense under projection he wrote: "b) Projection—Blaming own difficulties and inadequacies on environment—All my troubles are caused by way the faculty runs this school of psychiatry" to which the professor wrote, "Hmm!" (Bowen, n.d.a). Bowen received a "71 Satisfactory" grade on the test.

In October 1946, he was not selected for training in analysis (Bowen, 1948a, 1). It was not a setback in his mind. His reaction was "just give me time" (Bowen, 1948b).

Notable in Bowen's residency class notes are his frequent use of illustrations—diagrams, charts, graphs, columns and artistic renderings. While translating ideas into visual representations is a characteristic of qualitative research document analysis, Bowen's notations give clues to how he approached information to advance his own thinking. Figures 3.1 and 3.2 are examples of visually representing his speculations about mother/offspring/clinician transference relationships.

Looked at now, his ideas about family interactions between three members are clearly visible. Other points seen in the drawings are reciprocity in family relationships and transference substituting for change in the actual relationships. He was already thinking about the family as a unit in these drawings when he wrote, "What about father. He in here to extent M is not."

Bowen's orientations during the first two years were serious disorders such as schizophrenia and chronic alcoholism that were very resistant psychopathological processes within an individual. These drawings show his efforts to think beyond that to the connectedness-relationship process in psychosis and neurosis.

Speaking about families

In Bowen's first year, he explored a new treatment possibility.

> About 1 year after I became a psychiatrist [1946], I began my first efforts at seeing patients and family together. This was a setting which respected the sanctity of the therapist-patient relationship and the 'family' effort was 'smuggled under the counter.' It was done in evenings and on Sunday, and when the supervisor was not supervising closely. Since no one knew what to do with schizophrenia, there was a little more latitude in trying new approaches and some of this was 'in the open.'
>
> *Bowen, 1966, 2*

As early as January 1947, Bowen used a family diagram as a tool to understand a nuclear family. It was the same format he later used in the NIMH project, squares for males and circles for females (Bowen, 1947b). It served as a shorthand way of seeing a sizable amount of information about a family.

The next year, in May 1948, Bowen used the family diagram in a look at his own family (Bowen, 1948c).

The efforts Bowen made at family therapy during 1947 had unsatisfying results but this did not dampen his interest. This same year he formally presented his

WC #
234

N.D.

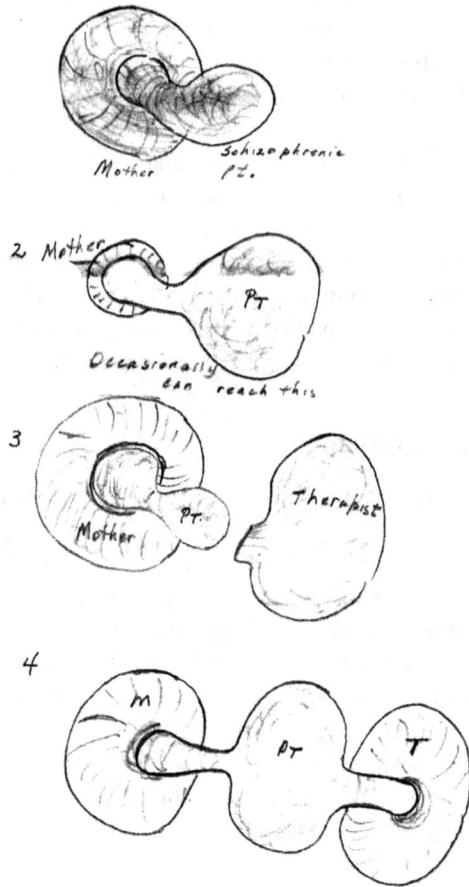

FIGURE 3.1 Bowen's class notes (Bowen, n.d.b)
Source: Reprinted courtesy of the Bowen family.

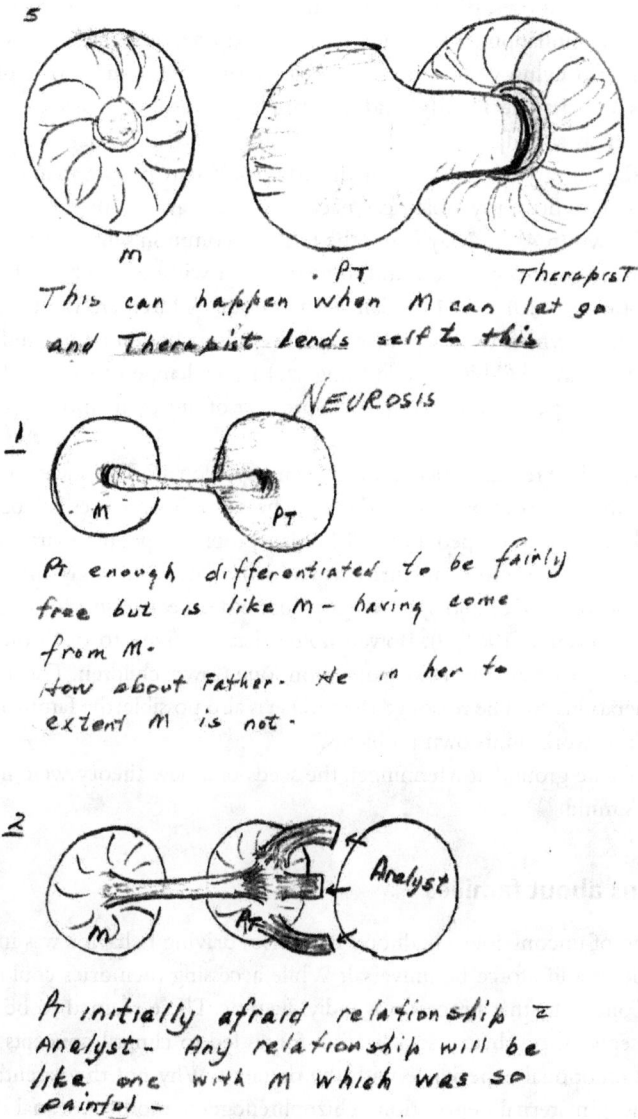

5

This can happen when M can let go
and Therapist lends self to this

NEUROSIS

PT enough differentiated to be fairly
free but is like M - having come
from M.
How about Father. He in her to
extent M is not.

PT initially afraid relationship =
Analyst - Any relationship will be
like one with M which was so
painful -

FIGURE 3.2 Bowen's class notes (Bowen, n.d.b)

Source: Reprinted courtesy of the Bowen family.

thinking about family in "Psychiatric Approach to Everyday Problems in the Family" (Bowen, 1947c). In this presentation, the breadth of his mindset comes through and it is possible to see the earliest underpinnings of a range of functioning.

The same mental mechanisms present in psychotic patients were also present in everyone, that mental illness was something of a matter of degree, and that neuroses, behavior problems, delinquency, crime, sexual offenders, divorces, unhappiness, national and international conflicts, and now effectiveness in living were all lesser manifestations of the same basic process .... **It is within the family that the most preventive work can be done** [emphasis added] (p. 3) .... The whole success of a marriage is determined by the emotional maturity of the parents. When people marry, they pick partners whose psychological needs in one way or another approximate their own (p. 4) .... Maybe use the term 'uncommon sense' for the ability to look at things objectively and to assess them without the bias of your own emotions (p. 10) .... More dangerous to our children are the things in our parents of which we disapproved. You were hurt by something and you over react by saying 'This terrible thing will never happen to my children' and we thus impart to our children the product of our own emotional struggles.

Was this straight Freudian theory or Bowen's spin on it? The ideas in the presentation remain as part of Bowen theory: The same basic forces of togetherness or individuality between people in all human problems; people marry spouses at the same level of emotional maturity as self, later amended to say this is "a clinical fact" (Bowen, 1958) and spouses have "opposite mechanisms for dealing with maturity" (Bowen, c. 1961, 9). Bowen noted that reactivity to one's own parents projects the process to the next generation, one's own children. This can repeat across generations. And he reasoned the reverse is also possible; the family is a source of strength to work on its own problems.

In the fertile ground at Menninger, the seeds of a new theory were incubating in Bowen's mind.

Questions about families

A life force of unconscious childhood memories driving behavior was individualistic. Wouldn't a life force be universal? While accessing memories could alter the course of one's life, that process was individualistic. These seemed to be two separate concepts. In psychiatry, socialization forces led to clinical concepts such as a corrective emotional experience with the therapist. Why not the parent? Parental aberrations of maternal deprivation, schizophrenogenic mothers, sexual deviations or long-term impact on a person from a single traumatic event were understood as causative of psychiatric problems. Yet, Bowen recognized plenty of similar examples, with no negative outcome within families and cultures. He had questions about siblings. How could one child in a family be impaired when other children from the same family were not? Parents with an impaired child also raised siblings with no symptoms. This led parents to feel blamed by professionals. Early on Bowen saw a schizophrenic patient whose parents were both professionals, father a psychoanalyst and mother a psychologist, yet the child was severely impaired. Pointing fingers was not the answer. More was occurring in these relationships than the intelligence to know how to raise a child without such problems (Bowen, 1986). Bowen knew there were families that had highly negative experiences and the children were normal (Bowen, 1986). Some way of understanding how this could happen was missing in psychiatry. The problem occurred with the application of Freudian tenets, on a case-by-case basis, by the professional, using, in the treatment outcome, a singular interpretation based on personal clinical judgment. While psychoanalytic theory defined emotional illness as an outcome of a process between parents and child in a single generation, Bowen considered it too deep to have occurred so quickly (Bowen, n.d.c).

Bowen questioned the exclusion of families from the treatment program. His interest in what comprised effective treatment and efficient treatment led him to novel approaches. Recognizing that the need for psychiatric consultation exceeded the time any one psychiatrist offered, he explored ways to make high-quality care available, from direct face-to-face sessions to phone calls to letters as consultation methods. This served both mental health purposes and his research purposes.

Clinical experience, doing the impossible, documenting the odyssey

Bowen had a positive regard for Freudian theory and PSA as a method, although to him neither the theory nor the application seemed broad enough to be grounded in the sciences. He wanted to make a contribution to the shift to a more scientific base. In 1988, Bowen reflected that going forward with his own thinking meant doing it alone:

> In the 1940s the profession used a strict interpretation of Freudian Theory. Important life forces were determined by infantile memories (unconscious) carefully hidden from conscious memory. A 'cure' was possible when these inner secrets could slowly be remembered with an understanding listener. It often required complete relaxation on a couch-month or years-for the patient to finally remember unconscious material. Unless the patient could trust the confidentiality of the analyst, the unconscious would not be revealed. The analyst was one who had been psychoanalyzed himself and who had spent years in PSA Institutes on the intricacies of theory. Impaired patients could not free associate (talk freely) to slowly remember the unconscious .... This meant the theory was right but treatment was not .... The new theory [Bowen's] was an individual pursuit in an institution devoted to proving Freudian theory by 'modifying' the activity of the analyst or therapist.
>
> *Bowen, 1988a*

That is exactly what Bowen did at Menninger: "modifying the activity of the analyst." The intent was to change the interaction and watch. Would new theoretical ideas about the human as a biological being develop?

Bowen's reflections in the 1980s of his odyssey enrich the history in the documents of the 1940s, giving us insight into his actions. Success in clinical experience on forming transferences with seriously impaired people, then considered impossible, and how to resolve transferences with this group, also considered impossible, led him to try avoiding forming them by relating in new ways to his patients. Almost 50 years later, Bowen reflected on the effort key to this:

> You absolutely cannot separate a self from any one person. It is resolvable in a triangle. I had a relationship, I did some research back in the 40s with these severe people, which it was possible to finally separate myself from them, but only after I got into family.
>
> *Bowen, 1983*

For Bowen, the accepted supposition—that transferences could not happen with the severely impaired, which meant that they could not be resolved either and supportive therapy was all clinicians could offer—was an assumption. He would set up situations and learn for himself the facts in the assumptions. For Bowen, family was the answer to the question about transference.

Questions persisted throughout the time at Menninger about what it would take for psychiatry to be a science. If medicine and its specialties can be scientific, how come psychiatry is not? Bowen was also remarkably consistent in time tracking his own odyssey as early as the late 1940s: "I have spent some thirty years working on the conceptual gulf between the social sciences and biological sciences ..." (Bowen, 1977).

It can be said Bowen came to Menninger oriented toward research. Comparing and contrasting course work with clinical work, he concluded that theoretical knowledge was incomplete. This was not criticism. It was opportunity. He was motivated to add more facts.

References*

*Unless otherwise specified, the works of Murray Bowen included in this chapter are from the Murray Bowen Papers. 1951–2004. Located in: Modern Manuscripts Collection, History of Medicine Division, National Library of Medicine, Bethesda, MD. The Accession, Box and Folder information differs for each one and that is included here.

Bourne, M. (1987). Accessing inherent resources in individuals and families. *American Journal of Family Therapy, 15*(1):75–77.
Bowen, M. (n.d.a). [Menninger test paper]. (Acc. 2007, Box 2, Folder Menninger).
Bowen, M. (n.d.b). [Drawing]. Copy in possession of Bowen family.
Bowen, M. (n.d.c). [Draft]. (Acc. 2005-055, Box 3, Folder Working papers).
Bowen, M. (1946a). [Letter on April 28, 1946]. Copy in possession of the Bowen family, Williamsburg, VA.
Bowen, M. (1946b). [Letter on September 8, 1946]. Copy in possession of the Bowen family, Williamsburg, VA.

Bowen, M. (1947a). [Letter, 1947]. Copy in possession of the Bowen family, Williamsburg, VA.

Bowen, M. (1947b). [Family diagram on patient at Menninger, January 1947]. Copy in possession of the Bowen family, Williamsburg, VA.

Bowen, M. (1947c). [Presentation. Psychiatric approach to the everyday problems in the family]. Copy in possession of the Bowen family, Williamsburg, VA.

Bowen, M. (1948a). [Letter on April 5, 1948]. Copy in possession of the Bowen family, Williamsburg, VA.

Bowen, M. (1948b). [Letter on January 15, 1948]. Copy in possession of the Bowen family, Williamsburg, VA.

Bowen, M. (1948c). [Family diagram]. Copy in possession of the Bowen family, Williamsburg, VA.

Bowen, M. (1958). [Emotional problems in older people]. (Acc. 2006-003, Box 7, Folder Family psychotherapy in families with a schizophrenic family member).

Bowen, M. (1959). [Letter to a former Menninger patient on July 6, 1959]. (Acc. 2003-044, Box 3, Name redacted).

Bowen, M. (c. 1961). [The origin and development of schizophrenia in the family]. (Acc. 2004-013, Box 4, Folder The origin and development of schizophrenia in the family).

Bowen, M. (1966). [Explanation of IDENTIFYING DATA section]. (Acc. 2009-013, Box 6, Folder GAP papers).

Bowen, M. (1973). [Letter on November 23, 1973]. (Acc. 2007-012, Box 2, Folder Misc. Correspondence).

Bowen, M. (1977). [Letter on August 17, 1977]. (Acc. 2004-013, Box 3, Folder Symposium 1977).

Bowen, M. (1983). [Transcription]. (Acc. 2007-012, Box 3, Folder 5/2/1983 First Year Post Grad. Class: Dialogue with Dr. Bowen and Class).

Bowen, M. (1984). [Letter on July 13, 1984]. Copy in possession of the Bowen family, Williamsburg, VA.

Bowen, M. (1985). [Presentation]. (Acc. 2003-044, Box 6, Folder Ortho – NYC April 24, 1985).

Bowen, M. (1986). [Letter to Bob Dysinger on October 12, 1986]. (2007-012, Box 2, Folder Partial drafts of letters).

Bowen, M. (1987). [Transcription, presentation at Georgetown Family Systems Symposia]. (Acc. 2004-013, Box 3, Folder Symposium-Nov. 7-8, 1987).

Bowen, M. (1988a). [Origins of fam. theory and therapy in mental health]. (Acc. 2006-003, Box 8, Folder Family evaluation).

Bowen, M. (1990). [April 19, 1990. Video clip. Addictions and family systems. A science of human behavior]. Retrieved July 20, 2020, from The Murray Bowen Archives: http://murraybowenarchives.org/videos/tape-one-part-one-a-science-of-human-behavior.

Firestein, S. (2012). *Ignorance: How it drives science*. New York: Oxford University Press.

Kansas Historical Society. (2004). *Menninger clinic*. www.kshs.org/kansapedia/menninger-clinic/12147.

Kerr, M., & Bowen, M. (1988). Epilogue: An odyssey toward science. In *Family evaluation*. New York: W.W. Norton.

Mackie, B. S. (2006). Early pioneers in the psychoanalytic treatment of psychosis in the institution. Presented at the 15th ISPS International congress for the psychotherapy of schizophrenia and other psychoses 12–16 June in Madrid, Spain, 1–13.

Mackie, B. S. (2016). *Treating people with psychosis in institutions: A psychoanalytic perspective*. New York: Karnac Books. Schlesinger, H. (2000). Musings about my intellectual development. Chapter in Part 1. P. Fonagy, R. Michels, & J. Sandler (Eds.), *Changing ideas in a changing world: The revolution in psychoanalysis: Essays in honour of Arnold Cooper*. London: Karnac Books.

Schlesinger, H. J. (2007). The treatment program at Menninger. *American Imago, 64*(2), 229–240. https://doi.org/10.1353/aim.2007.0027.

Smarsh, S. (2010). *It happened in Kansas: Remarkable events that shaped history*. Guilford, CT: Morris Book Publishing.

4

THE LAST YEAR OF THE RESIDENCY

A search for answers c. 1947–1948

The obvious discrepancies between his learning and clinical observations, combined with determination to help his patients, directed Bowen to an intensive literature search beginning around 1947. He was seeking information on how other disciplines used facts about the human. Had others wondered about human capacity as an extension of what's possible in nature? Life experience had already shown him that the possibility for human capability was more than is commonly known.

Bowen applied his expectations of psychiatry to be more factual to himself. Defining that template meant he could make identity comparisons with the organization and the field, and the trajectory of each. Defining personal principles helped reduce reactivity to others. An important principle was that the theory a psychiatrist held directed the therapy he gave his patient. Effective treatment required a solid theory, and a solid theory must be founded on an understanding of the origin of emotional problems. His meticulous search meant identifying discrepancies in psychoanalytic theory to avoid or to research further. "A thorough understanding of cause would provide reason for therapy" (Bowen, 1986a). Bowen's lifelong search was to be a more effective clinician. And that depended on the solidity and universality of the theory used.

He succinctly expressed his personal attitude as "Don't complain, don't adapt, [but do] assess, consider options, [and] make [my] own decisions" (Bowen, 1986a). There are hints he was not always well received when questioning his superiors and taking responsibility for his own learning. Following a path to discover the scientific underpinnings of psychiatry, then adapting Freudian theory treatment to his observations of his patients, predictably caused some at Menninger and, later, at National Institute of Mental Health (NIMH) to consider Bowen a maverick. His personal principles addressed both being an individual and managing pulls for togetherness.

DOI: 10.4324/9781003027287-5

Ironically, Bowen's approach matched the Menninger approach to help patients find their way. Reading, to help self, was part of the patients' treatment plans. Bibliotherapy, a time for reading, reflection and expanding one's own understandings was part of their standards of practice (Schlesinger, 2007, 231).

The literature review

Bowen's scholarly approach directed him to the rows of books available in the Menninger library. Because of the emphasis on the wisdom of great thinkers, the library was rich in books and other resources. Bowen spent time finding and "reading dozens of books that dealt with the human, or were about the human and science ..." (Bowen, n.d.a; Bowen (1987a). Reflecting the mind of a true systems thinker, he began exploring what theory ruled the disciplines of other professions. He read in the basic medical, clinical medical and nonmedical disciplines (Bowen, 1984a). The latter were interested in humankind's welfare but were not healers. He was asking what facts of human behavior are taught and practiced in each discipline.

Exploring the disciplines led to the conclusion that Freud's theory and its application in psychoanalysis was a baseline that society used to understand human behavior, despite its flaws. This generated the question of how to move Freudian theory toward science. Human progress in mastering the environment, forming cultural groups and producing advancements in language, arts, medicine and industry showed humanity's uniqueness in having a brain capable of such tasks. Thus, humans differ from any other living group, an idea aligning with Freudian theory (Bowen, 1986b). There are higher brain functions in the human. But there are similarities at the level of lower brain functions between the human and other living things, and a comprehensive theory must acknowledge those similarities. This focus was uniquely Bowen's and added substance to understanding human behavior. And human behavior is only the visible layer of human relationships. Behavior intersects with a human's physiology and sensitivity to interactions.

After reviewing disciplines, he studied how different sciences treated facts. He read in the physical (astronomy, mathematics) and natural (biology, anthropology, evolution) sciences, in social sciences and in subjective, interpretative areas, including philosophy and religion (Bowen, 1986c, 6). Papers in the Bowen archives reveal the contextual reading: the history of medicine as a science (biology) and as a healing profession (the art of medicine), the roots of sociology and psychology as they descended from philosophy and theology and of anthropology descending from evolution (Bowen, 1955). The first years at Menninger were a time of enlarging his mindset through literary explorations emphasizing the sciences. In Bowen's look back at this time, he noted that his survey of other disciplines produced significant changes within as he slowly advanced in the readings.

Bowen read in mathematics and physics, more factual disciplines that led to casual remarks with his contemporaries about whether mathematics or physics might lead to a more scientific base for a theory (Bowen, 1987a). Such remarks were a trial balloon on how ideas about theory and science might be accepted in that environment. After several years of investigation, this study did not introduce any workable indications (Bowen, 1987a). Astronomy and evolution readings, based on natural systems ideas, provided a solid "background blueprint" (Bowen, 1984b).

In Bowen's years at Menninger, he heard from the experts studying human behavior. American, European and South American specialists spoke to the residents. These included psychoanalysts Frieda Fromm-Reichmann and Margaret Mahler, anthropologist Margaret Mead, neurofeedback pioneer Elmer Green, ethologists Ludwig von Bertalanffy and Konrad Lorenz, and psychologist Jean Piaget, who gave seminars on genetic epistemology. Eric Erickson published his lectures at Menninger's in his 1950 book *Childhood and Society* (Schlesinger, 2007, 237–238). One can imagine the lively discussions these experts evoked.

Bowen's readings in these early years at Menninger led to the suggestion that the human was a phylogenetic progression from the lower forms of life. This thought had the potential to serve as a base for one all-encompassing theory as it integrated two theories, one that focused on nonhuman life and the other that studied how humans differ from other life. In the mid-1970s, Bowen reflected, it was in this extensive literature examination that a preliminary fundamental understanding occurred.

> In the 1940s, I became interested in the inconsistencies in conventional psychiatric theory. Reading led to the hunch that emotional illness is somehow related to that part of man that he shares with the lower forms of life, but that idea remained no more than a distant notion for the next decade.
>
> *Bowen, c. 1976, 1*

The only theory with facts that qualified regarding the natural world was, he found, evolution (Bowen, 1987b, 23). Conventional (Freudian) theory portrayed humans as unique, different from any other thing that ever lived. Bowen's awareness went to the other view of humankind as an evolutionary extension of the lower animal forms. This important nodal point that humans are more similar to than different from other animals germinated in the back of Bowen's mind for six more years. "I believed it would be helpful to treat man as directly related to all life, and to regard all the latter era developments as functions rather than things within themselves ..." (Bowen, 1982). This framing of functions is a critical component of his search. It had an influence on his thoughts and observations, though he did not have data for it yet.

Library committee

After numerous visits to the library and "bugging the librarian (Bowen, 1987a, 7) Bowen was invited to join the Library Committee. Bowen ended up being the one who ordered the books for the committee. This gave him latitude for his quest (Bowen, 1988). As a Library Committee member, he had access to the newest writings on Freudian thinking and its applications, and on scientific explorations then occurring. Bowen remained on the committee throughout his stay in Topeka (Bowen, 1987a, 8). In his selections, Bowen stayed with his guiding question, if psychoanalytic theory is not a science, what would it take for it to be a science?

Rudimentary new ideas from the readings

Pulling the idea of the family as a unit into coherence, giving backbone to a more factual theory awaited the clinical research at NIMH. During his years at Menninger he was "thinking toward this viewpoint. For instance, I was tending to view human courtship as governed more by rules that govern animal behavior than rules from man's sociocultural rulebook" (Bowen, n.d.c).

The phrase "natural system" does not occur in writings during the Menninger years. Yet Bowen operated as if that understanding motivated his approach. This understanding was congruent with the definitions of a natural system found elsewhere.

The forces of nature on living things form a natural system. It exists independent of, but includes, the human. A system formed by nature does not come from the mind of the human, though a person can observe a natural system and use tools such as mathematics to further understand it. For emotions, works about the human were not as clear as texts about animals. Writings had a vagueness, and there was a lack of discrimination between species. An essential difference in the literature showed two views of the human: one in evolution as an extension of all life, and the other as unique and separate from all other life on earth.

The psychoanalytic baseline of human uniqueness left out other facets in understanding human behavior. In Bowen's fact-finding readings through each discipline, he found support for a human emotional system.

> The literature refers to emotions as much more than states of contentment, agitation, fear, weeping, and laughing. It refers also to states in animals--including contentment after feeding, sleep, and mating, and agitation states in fight, flight and the search for food .... One might consider an emotional system present in all forms with an autonomic nervous system, but why exclude states of contentment and agitation in one-cell forms, in which stimuli could be more biochemical in nature. When one considers emotion on this level, it becomes synonymous with instinct which governs the life process of all living things.
>
> *Bowen, 1971, 198*

Darwin is not mentioned in this passage, but this is consistent with his work and the work of others who had similar ideas. Darwin's view that "science consists in grouping facts so that general laws or conclusions may be drawn from them" (Darwin, 1887, 48) was congruent with Bowen's path. In his explorations of other eminent scientists, Bowen found coherence between his reasoning and Darwin's work, which postulates a continuity between humans and animals in expressive behavior and emotional states.

> One man was most influential in helping me acquire the knowledge for a new theory. He was Charles Darwin, who was clearest in his writings about evolution. Einstein contributed fantastic knowledge about the space between people. Then came the knowledge about DNA.
>
> *Bowen, 1981*

Human behavior is complex. The systems readings he did of models from mathematics, though popular, and before biostatistics, were lacking the harmony within nature of the human as a life form (Bowen, 1987c). Bowen recognized the essential need for a connection between evolution and its many variables within the breadth of human behavior.

Bowen, when preparing a chapter for the 1988 book *Family Evaluation*, wrote out the laborious process that began in his residency for discovering connections between human behavior and evolution. It is a formula anyone can use.

> It involved a continuing process of cross checking everything learned in the extensive training courses, with experiences from clinical practice, with public and private communication from senior staff and instructors, with all the books in the library. If one could list the differences (in the books), the number would be overwhelming. If one eliminated statements based on feelings, imagination, intuition, mythology and other items of subjectivity, the list moves a little closer to objectivity, and it is more manageable. If one eliminated the feeling items and focused on common denominators, instead of differences, the list becomes manageable. The goal included an evaluation of the origin and content of each professional discipline. It was fortunate to try to separate the feeling-subjective states from the objective (non-feeling)–scientific states .... Out of the amalgam came some clear ideas about what had happened with psa [psychoanalytic] theory, followed by some ideas about how the pitfall could be avoided.
>
> *Bowen, 1986a*

One can imagine the challenging discussions that took place at those library committee meetings. His relationship with Karl Menninger and the library experience was a permanent gift to the future of Bowen's theory development. In a letter of appreciation to Menninger, Bowen confirms this.

> You were one of the main architects of my early life 'compass.' The compass wondered what would happen if Freud and Darwin were combined. It took a natural systems theory to handle the variables, but the compass knew the

way when feelings suggested a detour. The long term library committee …
played a contributing role.

Bowen, 1988

Bowen's beliefs

From his efforts at discussion with mentors and peers, Bowen also knew that
"ANIMAL IN MAN" [caps in original] (Bowen, 1973) was an idea out of place in
his time. Recollecting the challenge he faced, Bowen wrote, "This idea, so radical at
the time, it was impossible for me to comprehend. [I] could find no clues to implement these ideas and they remained in the far distant background for a decade"
(Bowen, 1977). It was in 1956, when an understanding of the family as a unit was
in place, that Bowen had enough information to set up a search directly exploring
such a hypothesis.

His belief that "emotional illness relates to that part of the human shared with
the lower forms of life" (Bowen, 1981) was a true search. Little existed in the literature on humankind as a part of all life compared to the wealth of writing on
how humankind is unique. The search centered on unambiguous, direct theoretical connections, not just on analogies. Humankind as connected to all of life
strengthened into a long-term hypothesis about the human as part of a natural
living system and required a theory. Bowen reflected on his mindset before going
to NIMH:

> I already had a fair concept of natural systems from early life experiences, from
> voluminous reading in the natural sciences during the 1940's, and decades
> of pondering the ramifications of natural systems to the social sciences, all
> before I started family research.
>
> *Bowen, n.d.c*

Bowen's principles had important consequences for how a theoretical model
formed. In considering the human an instinctual biological being, a part of all life,
he excluded anything derived from literature, the arts and the inanimate world. The
theoretical model had to be consistent with human beings as a part of nature and
Bowen's personal effort would go "to think toward the biological sciences" (Bowen,
1976, 3). He had full awareness such efforts to solve this theoretical problem went
beyond a lifetime's work.

A deeper survey of the beliefs Bowen arrived with to Menninger remains for
another interested explorer accessing the archives.

Beginning the shift to a new theory

The third and final year of his residency brought a choice, to stay at Menninger
or to step out into the world of psychiatry elsewhere. Bowen was ambitious and
forward thinking. He had already made choices that were not based on money

and now had choices to make based on what was best for his interests. Other places were contacting him—Kansas City, Yale—and repeated inquiries from Brill to come to Washington, DC and practice with him. While financial return was a consideration, it was not primary. Bowen had established that in choosing to go to Menninger, the premier place for a residency (Bowen, 1948a, 2). As he headed into 1948, he began concentrated efforts at applying his new ideas to schizophrenia. The residency at Menninger required Bowen to apply Freudian theory in vivo, with the emphasis then on intensive individual psychotherapy for hospitalized schizophrenic patients (Bowen, 1966, 27).

The School of Psychiatry was an experiment within itself in its efforts to apply psychoanalytic theory to hospitalized patients via a milieu designed to do that. Bowen had advanced there. His perception of those in authority was "the attitude of the Menninger leaders were more interested in helping young people develop their own capacities than in communicating a fixed body of knowledge" (Kerr & Bowen, 1988, 349). He stayed at Menninger. He had ideas he wanted to explore using regression therapy. This would be an intense clinical exploration, becoming the prompt that would lead to two new concepts and a prototype for clinical practice.

Bowen was making a name for himself in psychiatry. In January 1948, he accepted a position as consultant, a junior staff member position, at Winter Hospital, where he had supervisory and teaching responsibilities for VA trainees (Bowen, 1948a, 1). This same month, Bowen's organizational skills offered him advancement to Assistant Medical director of the Menninger sanitarium. This put him in a supervisory position to other residents (Bowen, 1948b, 1). It was a year where his future choices were on his mind. And he had no problem thinking big on possibilities of future advancement at Menninger. "I have a fairly high batting average and more than average chances to run the main show" (Bowen, 1948a, 2).

In reviewing patient progress notes, a duty of being a supervisor at Winter Hospital and Menninger, Bowen encountered curious, contradictory notes written on patients who had been in the hospital five years or more. People stayed at Menninger that length of time, and the hospital required a note on progress to be written every ten days. In a five-year period, that obligation equaled 180 notes. The notes said the patient was getting better, but why was the patient still hospitalized? How did the family fit in here? Then current hypothesis pointed to those relationships as the origination of the problem. And observations showed that improvement declined after a home visit. The family was a missing piece in the treatment plan. This experience tapped into Bowen's efforts to understand what constitutes progress and to set his own metrics. Bowen heard the reports of clinical successes found in the charts often told at lunch or coffee breaks by the writers of those notes. How was it the patient improved, yet he or she remained hospitalized after these many years? Bowen began paying attention to these notes, comments and conversations, discovering that reports of improvement came when there was a high positive regard between the professional and patient or, contrarily, when the professional figured out a manipulation by a patient and responded strategically

(Bowen, n.d.b). A note of progress depended on the relationship between the person and the professional writing it. "Observations were determined by 'the mind-fix' of the observer rather than the nature of the situation to be observed" (Bowen, n.d.a).

Bowen wanted a way of objectively reporting a patient's status. Wouldn't a measure of solid change involve more than opinion? There was nothing on maintaining improvement once living with family again. Improvement rarely seemed to transfer from clinician to family.

As far back as his Army research, Bowen oriented to the essential qualities of the people to be observed in an environment. This was the basis for his neutrality, observing even the smallest interactions as possibilities to be explored in his orientation to a theory-based practice. This foundation for his explorations moved him to what he could research. His scientific method was one of reading, searching for facts, observing, trying new applications (not to confirm them but to name more of what was possible to be known), seeking patterns and repetitions, and continuing to extend the questions. Convincing himself, moving on to new understandings, to more substantive findings, advancing the overall effort toward a more accurate knowledge of human behavior meant leaving many details to be filled in by later researchers. He had the conviction that the facts of any theory must stand the test of time offered by the choice made in applying it.

Bowen had a deep resolve that a theory based in science could explain human behavior and offer a guide to corrective action for humankind to have a better life (Bowen, 1966). "Research allows for the possibility of prevention which benefits all of mankind" (Bowen, 1956). He did the years of work, exploration and study essential to move toward this possibility. Bowen's approach to the question of why psychoanalysis was not a science was to learn everything possible about the theory. The very fabric of the Menninger School of Psychiatry was to offer proof of the theory. Bowen went beyond this, systematically strengthening the theory by replacing assumptions with facts.

The march to a new theory had begun.

References*

*Unless otherwise specified, the works of Murray Bowen included in this chapter are from the Murray Bowen Papers. 1951–2004. Located in: Modern Manuscripts Collection, History of Medicine Division, National Library of Medicine, Bethesda, MD. The Accession, Box and Folder information differs for each one and that is included here.

Bowen, M. (n.d.a). [Presentation on what family therapy cannot do]. (Acc. 2005-055, Box 3, Folder Working paper – re: Presentation on what family therapy cannot do).
Bowen, M. (n.d.b). [Letter]. (Acc. 2007-012, Box 1, Folder misc. working papers).
Bowen, M. (n.d.c). [Draft of a letter]. (Acc. 2007-012, Box 3, Folder Bowen position statement/letters).
Bowen, M. (1948a). [Letter on January 15, 1948]. Copy in possession of the Bowen family, Williamsburg, VA.
Bowen, M. (1948b). [Letter on March 28, 1948]. Copy in possession of the Bowen family.

Bowen, M. (1955). [The current status of man in relation to mental health]. (Acc. 2006-003, Box 5, Folder Mtg. notes October 1954).

Bowen, M. (1956). [Letter on September 14, 1956 to a co-worker of someone who knew Bowen]. (Acc. 2006-003, Box 6, Folder NIMH correspondence H through P).

Bowen, M. (1966). [Explanation of IDENTIFYING DATA section]. (Acc. 2009-013, Box 6, Folder GAP papers).

Bowen, M. (1971). Family therapy and family group therapy. Chapter 10 in *Family therapy in clinical practice*. New York, NY: Jason Aronson, 1978.

Bowen, M. (1973). [Letter on November 23, 1973]. (Acc. 2007-012, Box 2, Folder misc. correspondence).

Bowen, M. (c. 1976). [Letter]. (Acc. 2005-055, Box 3, Folder Murray Bowen M.D. position statements).

Bowen, M. (1976). [Psychiatric contribution to study of the family]. (Acc. 2003-044, Box 4, Folder Smithsonian – June 1976).

Bowen, M. (1977). [Notes for presentation at symposium, 1977]. (Acc. 2004-013, Box 3, Folder Symposium 1977).

Bowen, M. (1981). [Letter on September 18, 1981]. (Acc. 2007-012, Box 2, Folder Bowen position statement/letters).

Bowen, M. (1982). [Subjectivity, *Homo sapiens*, and science]. (Acc. 2004, 013, Box 3, Folder Pre-symposium, Nov. 14, 1982).

Bowen, M. (1984a). [Letter to conference organizer]. (Acc. 2003-044, Box 5, Folder name redacted – Chicago, IL, June 14–16, 1984).

Bowen, M. (1984b). [Presentation March 16, 1984, Northampton, MA]. (Acc. 2003-044, Box 5, Folder Northampton, Mass – Mar. 16, 1984).

Bowen, M. (1986a). [Draft]. (Acc. 2006-003, Box 8, Folder Drafts of 'odyssey' – In theoretical principle).

Bowen, M. (1986b). [Presentation notes]. (Acc. 2004-013, Box 3, Symposium – 23rd, Nov. 8–9, 1986).

Bowen, M. (1986c). [The place of family in the future of the behavioral sciences]. (Acc. 2003-044, Box 6, Folder Del Ray Beach Florida, July 25–26, 1986).

Bowen, M. (1987a). [Draft]. (Acc. 2006-003, Box 8, Folder printed papers in odyssey, Part 1 of 2).

Bowen, M. (1987b). [Transcription, presentation at Georgetown family systems symposia, November 8, 1987]. (Acc. 2004-013, Box 3, Folder Symposium-Nov. 7–8, 1987).

Bowen, M. (1987c). [Draft]. (Acc. 2006-003, Box 8, Folder printed papers in odyssey, Part 1 of 2).

Bowen, M. (1988). [Letter to Karl Menninger on August 27, 1988]. (Acc. 2017-012, Box 3, Folder correspondence between Karl Menninger and Murray Bowen).

Darwin, C. (1887). Autobiography. In F. Darwin (Ed.), *The life and letters of Charles Darwin: Including an autobiographical chapter*. John Murray. https://doi.org/10.5962/bhl.title.1416.

Kerr, M., & Bowen, M. (1988). Epilogue: An odyssey toward science. In *Family evaluation*. New York, NY: W.W. Norton and Company.

Schlesinger, H. J. (2007). The treatment program at Menninger. *American Imago, 64*(2), 229–240. doi:10.1353/aim.2007.0027.

5

MODIFYING ACCEPTED PSYCHIATRIC THEORIES AND PRACTICES 1948–1951

The years 1948–1951 were pivotal for Bowen's maturing orientation to family. In the Freudian-oriented Menninger clinic, the residents' priority was to learn psychoanalysis. But Bowen's priority was to study and investigate its base in science. This chapter gives an overview of Bowen's in-patient studies at Menninger. My intent is to describe the in- and out-patient explorations as separate segments. In reality, these explorations were overlapping; ideas from one effort influenced other efforts in his continuous search for a scientific base to Freudian theory. My reflection here is that the experiments supplied the skeleton for an embryonic theory of natural systems to materialize.

In the last year of his residency, 1948, Bowen's ability to operate in various spheres, as a clinician, a ward physician, an independent thinker or a representative of Menninger in the community, became most visible. It is the year he first experimented with psychoanalytic theory and therapy in his clinical work and began personal psychoanalysis (Bowen, 1963). One conclusion he drew from these activities was that psychoanalysis was not yet science, though the two-person relationship system Freud had developed using a therapist and a patient was a start.

Learning about families

Bowen interviewed relatives during their periodic visits with the hospitalized family member in 1948, which generated data first-hand. In contrast to the rejecting-mother proposition, during Bowen's informal family meetings, he observed intense attachment between mother and child. While a process occurring between the two people was consistent in these contrasting views, how Bowen understood the process differed. He did not interpret the behavior. Moreover, the reality of his meetings with family members did not support the prevailing belief that contact between

DOI: 10.4324/9781003027287-6

therapist and family contaminates the patient/therapist relationship. Rather, contact gave a deeper understanding of the mother and child relationship.

His direct observations also showed that recommending psychotherapy for the relatives could backfire. They withdrew from a relationship with the psychiatrist or pulled the family member out of treatment at such a suggestion. Bowen changed the approach to families, using a circular method to observe relationships, learn from the observations, apply the learning in novel ways, observe relationships and so on. By recognizing the family's sensitivity to any hint they had contributed to having such an ill child, Bowen opened lines of dialog to acknowledge the reality of other family problems. This is interesting because it shows that the family's reactivity to being blamed matched the blaming done by professionals, such as their introducing the phrase "schizophrenogenic mother."

> [A] schizophrenic is painfully distrustful and resentful of other people, due to the severe early warp and rejection he encountered in important people of his infancy and childhood, as a rule, mainly in a schizophrenogenic mother.
>
> *Fromm-Reichmann, 1948*

Increasing resources for families

To experiment with avoiding the family feeling blamed, Bowen tried a unique approach. He changed the milieu by altering how staff functioned. The ward administrator, responsible for the smooth running of the ward within the hospital, became available for day-to-day questions from the families.

> A ward [administrator] was available and he was not involved in the psychotherapy. This was distinctly the addition of a person who was not identified with the patient's problem. The overall attitude was that the illness was confined within the patient, the treatment program was oriented around the patient … and the attention to relatives seemed more to permit the patients' treatment to go better.
>
> *Bowen, 1957a*

The intent of this design was to de-stress families, precluding the removal of a patient from the hospital due to some upset and the hospital's use of legal means to prevent this (Bowen, 1956a). Bowen also offered social work services beyond intake assessments directly to families so they could speak more freely in private and have continuous active contact with the social worker. Involving social workers with families was an accepted way in psychiatry for handling transference. This kept sacrosanct the psychiatrist/patient relationship, though Bowen was unconcerned about contaminating his relationship with the family. Rather, he adapted this method as an adjunct to reducing family anxiety around blame.

This design offered two people to families not directly involved with the patient's treatment. The ward administrator helped to resolve issues between the families and the hospital and the social worker gave families a person to speak to about family concerns beyond the ill family member.

> This was a most significant change ....The addition of a social worker went beyond the provision of a person to relate to the relatives .... It was recognition that the relatives too have problems .... Some relatives had little more than passing acquaintance with the social workers. Others formed rather intense relationships. Some of the better clinical results were with patients whose relatives had the more prolonged and more intense relationships .... In retrospect, the importance of this stage seemed to be a recognition that the relative needed someone for his own problem and the addition of the social worker attempts to supply such a figure.
>
> *Bowen, 1957a*

Besides being better supported, the family now had first-hand information on what was actually occurring after the patient's admission. The family could voice their ideas about what would improve their situation. And the social work meetings gathered evidence of the family's adequacy and willingness to offer a supporting relationship to the patient. This was a two-pronged approach: intensive individual psychotherapy with the schizophrenic offspring along with casework with multiple members of the same family (Bowen, 1965, 7). The recognition that families were a resource to the professionals and integral to the treatment plan was one of the earliest paths toward family psychotherapy (Bowen, 1965).

But this approach revealed a problem. The collaboration between the social worker and the therapist highlighted that each had a separate perception of family problems from their contacts, and neither had the completeness of information that seeing the entire family provided. In this collaboration, "... strain with intense intrafamily issues [was] showing up between the therapist and social worker" (Bowen, 1956a).

This observation foreshadowed what occurred in the first year of the National Institute of Mental Health (NIMH) project. Family members' tensions could transfer to tensions between the social worker and the psychiatrist or between staff nurses; in turn, relationships in the family calmed. The important piece was that after years of working to contain upset between a twosome, introducing the social worker revealed evidence of a process between three persons.

Bowen went further than adding supportive staff for families. Family visits themselves became part of the treatment plan (Bowen, 1957a). They were prescribed. This had an important clinical outcome. The patients did better when families had casework support, and the families were more active in the overall treatment process.

As described in Chapter III, the first model Bowen used with families was direct contact by letter or by observing their visits. The second model encouraged contact between social worker and family and between psychiatrist and patient but not cross-therapeutic contact between social worker/patient and psychiatrist/family.

Neither psychiatrist nor social worker could confirm the other's report as there were no direct observations by the social worker of the patient or by the psychiatrist of the family. These reports depended on the professional's ability to accurately convey what each observed.

However, this second model engaged the family's cooperation differently. Family involvement and responsibility for their own family problems came together. No longer was the hospital in the position of reactive rule-maker to separate patient and family when disagreements arose. Each side had someone to consult. These early models still considered the patients as sick (Bowen, 1956a). In this model, patient upset was reduced because the home environment and family relations were improved. Bowen showed that positive family relationships enhanced therapeutic response. Bowen thought that a family's health existed partly in their attitude and supports like this eased family tensions to bring more positivity to their attitudes.

Bowen's personal psychoanalysis

As Bowen was more and more actively engaging with and learning from families, he began his own psychoanalysis with Ellen Simon in early April 1948 (Bowen, 1948a). It ended in February 1950, when Simon moved to New York City (Bowen, 1950a, 3). At the end of his analysis, Bowen noted it cost more in money, time, trepidation, angst and study "than my MD. cost, but it is the soundest investment I have ever made" (Bowen, 1949). The relationship between Bowen and Simon continued and the archives contain correspondence between them at least through 1959, when Bowen reviewed with Simon his plans to leave NIMH (Bowen, 1959).

After beginning analysis, considering his own family was a natural outgrowth (Bowen, 1950b), and he reflected on it nearly 30 years later.

> As an oldest son and physician, I had long been the wise expert preaching to the unenlightened, even when it was done in the guise of expressing an opinion or giving advice .... During my psychoanalysis, there was enough emotional pressure to engage my parents [others] in an angry confrontation .... At the time, I considered these confrontations to be emotional emancipation. There may have been some short-term gain ... but the long-term result was an intensification of previous patterns.
>
> *Bowen, 1972, 484*

He put familial engagements of this sort on hold when the application of psychoanalytic understandings with his family produced no solid progress. But when the impression of the family as a unit formed from research observations of the NIMH families, he returned to studying his own family (Bowen, 1978, XIV).

The interactions of both his own family and the clinical families were open to the research questions in Bowen's mind. For more breadth, in January 1947, Bowen used a family diagram for his studies (Bowen, 1947), the first known instance of using this tool. Figure 5.1 shows his hand-drawn clinical family diagram, done in

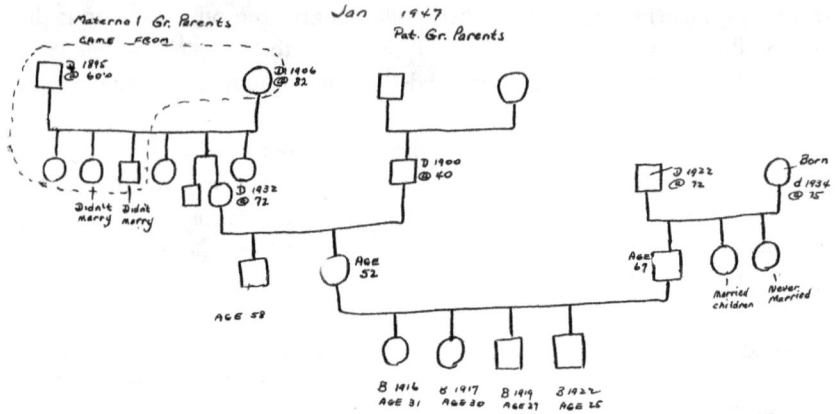

FIGURE 5.1 Clinical family diagram (Bowen, 1947)

Source: Reprinted courtesy of the Bowen family.

January 1947, of four generations on the maternal side and two generations on the paternal side.

The format of the diagram accounted for details of import in the family and in the connections to its past. By May 1948, while starting his personal analysis, he was using this tool to look at his own family (Bowen, 1948a).

Speaking to the local community

In the April that Bowen began his analysis, he presented "What can a Local Rotary Club do in Regard to the Mental Health Problem" (Bowen, 1948b) to the District Rotary Club Convention, in Topeka, KS. This was one month before Will Menninger sought to bring an awareness of mental health to Harry Truman, then running for president, and the same year Menninger published his book on state sanitarium reform. Interestingly, in Bowen's presentation, he spoke in the plural, which raises the question, Was he expecting to help Dr. Will's efforts or was Bowen speaking for himself and other residents? In this presentation, there is a seed of his future concept of a differentiation of self scale when he described the changing attitude toward mental patients:

> We learned that patients are made and not born, that the same personality defects and struggles present in them were present in all of us, that mental illness was more a matter of degree, and that personality difficulties, all the way from the severe psychoses through the hysterias, the neuroses, criminality, divorce, sex crimes, delinquency, alcoholism, racial prejudice, individual and group conflict, all the way down to simple human unhappiness and inadequacy—were not separate and individual problems but that they were

merely different manifestations of the same basic personality struggle. This is a different concept from the old one about mental illness.

Bowen, 1948b, 1

His rotary club presentation agreed with Menninger milieu practices and the political shift from warehousing people deemed mentally ill to prioritizing treatment. Bowen's interest in societal process began at least by the 1940s (Bowen, 1974, 269).

Observations on the ward

Back at Menninger's, Bowen was the Hospital Physician, also known as the ward psychiatrist, which meant that he evaluated patient's progress through daily meetings with them and through contact with hospital services such as nursing, occupational therapy (OT) and recreational therapy (RT) (Schlesinger, 2007, 234). He also saw patients who asked for psychotherapy. The meetings, along with Menninger's openness to and encouragement of serious consideration of therapeutic explorations, led Bowen, by October 1948, to report to his family that he was on to a new and significant approach.

> I have developed a new approach to schizophrenia, which I think holds some of the answers that have eluded psychiatry for so long. I have gone into this with all the energy I have and so far have just a little support for my ideas. Most of the people will accept some of my ideas, but they still stick to the general premises of the old ideas and will not go all the way.
>
> *Bowen, 1948c, 1*

The premise in psychoanalytic theory that the mother was "'bad' for the" child (Bowen, n.d.a, 3) was dissonant with nature. The new approach was his effort to make sense of the dissonance. The method was "… a form of anaclitic therapy … started for a series of severely impaired patients" (Bowen, 1956b), but it was informal research (meaning the Menninger research department did not sanction it).

Bowen's regression efforts took place on the ward at the Menninger sanitarium and at Winter Veteran's Hospital, which treated patients with schizophrenia and patients with alcoholism. What is important is the two locations had different milieus and staff. He began devising an in-patient anaclitic program based on observations and information gained through the social worker's contacts with patients and their families in 1949. Over the years of using regression (there were at least four iterations, 1948–1952), the subjects were people with either chronic schizophrenia or alcoholism.[1] He used this program with the two different diagnoses to broaden his study (Bowen, 1948d). Two of these patients are discussed in Chapter 6.

The regression candidates were understood to be unable to form transferences. They were contenders for supportive therapy only (Ruffalo, 2018) and Bowen, as a hospital physician, would have been the provider. He considered this assumption ripe for exploring and believed more would be learned about an inability to form transferences if treatment was investigated at the gross level of functioning (Bowen, 1957a).

Anaclitic or regression therapy overview

American psychoanalysts, beginning with Sullivan (1927, 1931) and Levy (1931), believed that schizophrenia was an outcome of rejecting/inadequate/overprotective mothering of the child. Anaclitic treatment or regression therapy offered a corrective emotional experience. Mical Raz, an Israeli researcher interested in the history of medicine, defines anaclitic:

> The term anaclisis (deriving from the Greek, 'to lean on') was first introduced into psychoanalytic theory as a translation of Freud's usage of the term *Anlehnung*. The term designates the early relationship of the sexual instincts, which at first are dependent on organic sources, to the self-preservative instincts. Hence, an anaclitic type of object choice is that which is based on the object of the self-preservative instincts, usually an object serving a parental figure.
>
> *Raz, 2010, 55*

Claude Nichols and Bernard Bressler, who conducted a study on anaclitic therapy in 1958, defined the treatment this way:

> Anaclitic therapy is an organized approach to the problem of a patient with a severe psychological disorder. This therapeutic approach utilizes the dependency state in the patient in promoting an orderly, planned program of regression with a nurse as the central therapeutic figure. The nurse who ministers to those dependency needs, who gives the patient unconditional physical and emotional support, who gives him care as a mother does an infant, promotes the regression phenomenon which is the cornerstone of this therapeutic approach. Anaclitic therapy may accordingly be defined as treatment in which the positive feelings and behavior of the patient toward the therapist (reminiscent of the infant's relationship toward its mother) are utilized to bring about psycho-physiological homeostasis.
>
> *Nichols & Bressler, 1958, 989*

Some practitioners of anaclitic therapy and direct analysis of schizophrenia, such as Marguerite Sechehaye and John Rosen, were guest lecturers at Menninger during Bowen's time there. Bowen had read Sechehaye's reports on her work (Bowen, 1951, 17).

Bowen described anaclitic therapy as the "substitution of one instinct for another" (Bowen, n.d.b, 3). The essential, natural, life-giving closeness between a mother and her offspring was replicated in a treatment relationship. In a 1957 look back on his work, he reviewed the various iterations of treatment methods he tried at Menninger before the change to regression.

> The treatment goal … was that the 'child' become an adult … 'How was the relationship to reach such a resolution?' and 'Who was to take the active role?' … The therapist might lecture the patient, apply subtle pressures, or otherwise urge the patient toward growing away from the therapist. Experience taught … that this procedure usually resulted in a more infantile and clinging patient. Another method, having a prototype in animal studies and certain human cultures, was for the therapist to set a date at which he terminated the relationship. This method had both advantages and disadvantages, and it was much less effective than a new method which was developed. In this approach the therapist assumed an active role in a multiple-stage weaning process. The usefulness and effectiveness of this method depended on the therapist's achieving a workable level of objectivity …. Patients had long experience in perceiving themselves as helpless and in presenting themselves as a helpless child; the therapist had long experience in accepting the infancy states of the patients as a reality.
>
> *Bowen, c. 1957a*

Overview of Bowen's approaches to using regression therapy

Bowen wrote that his attempt at standard regression, psychiatrist and patient, had unsatisfying results (Bowen, 1957b). This section reviews how the therapy evolved as Bowen searched for the most effective method. Chapter 6 will examine specific uses of regression therapy.

The late 1940s saw him experimenting with the common assumptions in Freudian theory (Bowen, 1962, 2–3). The point was to have the patient regress to an approximation of the preverbal level, an attempt to access a specific level of brain development consistent with the regression level goal of functioning. This idea came from analysts' observation of the difficulties patients had of putting their conflicts into words (Bowen, 1953, 4). John Rosen, who promoted direct analysis using this method, wrote that "schizophrenia is a disease which has its inception somewhere between birth and prior to the termination of the pre-verbal period and is caused by the mother's inability to love her child" (Rosen, 1953, 99). Bowen's effort was to replicate mothering attention that was missing in early life experience. There is no evidence that Bowen accepted the disease or defect concept of schizophrenia, but he did consider it a phenomenon that was too deep to have occurred in only one generation.

Bowen modified anaclitic treatment with selected, very ill in-patients who had families living at a distance. The theoretical hypothesis was to supply the missing maternal attention via nurses. This would facilitate innate growth within the individual, who then would grow up on their own (Bowen, 1957b, 1).

Rather than getting to unconscious memories or primary process materials, regression therapy was to provide good mothering by nurses and informal supportive therapy by Bowen in his position as a hospital physician (Bowen, 1956b, 1–2). The theoretical goal was to give as much attention and time as the patient could accept from the nurses (Bowen, 1995, 24). The patient could have many demands, mirroring the mothering observed in early developmental stages of the human and other species. The mother serves the infant, gratifying every need, which fosters the infant's innate growth and development. When every need and want was within the reach of being met in an adult patient and was being given with no effort from the individual patient, regression would occur. Then the individual would activate resources within self for managing the environment.

Some staff balked at what seemed contradictory to good medical care. Fostering regression with no encouragement to change was not considered good care. Bowen modified the goal from assisting regression to explicitly assisting growth. Emphasizing conditions to spark the full process from regression to growth enabled the nurses to find positivity in the regression, and the nurses found this more acceptable.

Helplessness in the person was relationship dependent, and the nurses were to give freely. It was important not to usurp the patient's choices or the inborn ability to identify what he or she wanted. The patient directed treatment. An important basic principle of this early work was the prediction that a person's innate growth, the life force of differentiation that makes each person unique, would be activated into becoming a separate individual.

Bowen discovered that leaving the activation of growth within the person required a paradoxical approach. Rather than encouraging more effort from the person who showed a weak urge toward mature activity, the staff encouraged the person to, for now, "relax, take it easy" (Bowen, 1957a), which would tap into the life force toward individuality and reinforce the inborn "I can exist on my own." But these ministrations did not result in the expected growth.

An important modification

The next iteration involved anaclitic therapy with both the psychiatrist and nurse functioning in a parental role while also extending social work services to the parents. In the pseudo parental in-patient effort, the caregivers—Bowen and the nurse—replicated family roles in the artificial environment with an intent to create a setting where the treatment they provided would be accepted (Bowen, 1962, 4). In this modified version, the patient would accept "love" in the form of having their demands met as much as possible by the caretakers (Bowen, 1957a). There are interesting notes in the archival papers of Bowen driving around Topeka to find pasteurized apple cider because the individual had requested it (Bowen, 1951, 17–18). Another patient, along with their special nurse, accompanied the Bowen family on vacation to Colorado (Bowen, 1987). In another note, Bowen documents walking with a patient to meet the patient's father, while the patient handled saved feces in his pocket (Bowen, 1951). A theoretical premise held that when staff met the person's every request, innate "urges and motivation," understood as "more mature activity," would emerge (Bowen, 1962, 4).

The influence of observations

Bowen had now applied two methods, one with the nurse as a primary care-taker and one with Bowen and the nurse in caretaker positions. Direct observation of a mother and son brought the next iteration, implemented in 1951, to attend a patient and eliminate any demands on them from caretakers. In Bowen's direct observations of schizophrenic patients with their families (Bowen, n.d.a, 2), a mother gave excessive love (smother love) to her child, but there was more to it. She could not also give love without making demands. Contrary to withholding caring, then compensating with excessive love as "defensive denial" (Bowen, 1962, 2), this was giving love but creating dependency with the bind of inherent demands.

> Reasoning said that if the problem was a deprivation of love, then the observation did not fit with this basic premise. Could it be the excessive attention or the demand within the giving that was a problem? Could the treatment program be shortened with the creation of a situation in which it was possible for the patient to accept love and attention and for the environment to give as much the patient could accept without any demands on the patient?
>
> *Bowen, 1957a*

Bowen devised a way to explore the promotion of psychological growth while testing the theoretical assumptions in psychiatry of maternal deprivation as a contributing cause of both chronic schizophrenia and alcoholism and inability to form a transference. An inability to form a transference showed a deficit in ability to form a bond with someone. "In essence, this difficulty signifies a lowered capacity to develop feelings and relationships with other people" (Weisbart, 2014, 1682). Mothering attention and hospital services were made available to the patient to choose or reject. This encouraged regression. The patient, who had been used to receiving caretaking with suggestions or encouragement, now had no expectations made of them.

> A treatment program was set up to go as near 'all the way' as humanly possible in providing the kind of attention a mother provides a child … to get by the rejection barrier in the patient. Full-time special duty nurses were … to provide any attention acceptable to the patient … the treatment goal was the opposite of the usual hospital milieu program. Staff demands, expectations, criticism and interpretations were reduced to the lowest minimum. It was expected that the patients would regress. This helps the giving of attention. If the patient regressed to an infantile level and could accept bottle feedings,

this too would be provided. The program was geared to continue indefinitely or until the patient showed signs of growing up on his own.

Bowen, c. 1957b, 6

This was not a treatment effort based on hourly schedules. Staff were available around the clock to meet every demand for service or attention. Bowen hired special duty nurses, or they were assigned to make this degree of attention possible. The involved staff changed their normal caretaking responses by minimizing interpretations, and by reducing encouragement and demands for improved functioning. Encouraging growth was another demand. The staff, therefore, presented a "no rush to grow up" response that gave latitude for the patient's own growth to take hold. Being present without judgment was a naturalistic approach. Bowen noted that much of his time in staff discussion was spent on understanding the patient, the treatment goals and supporting the giving "without hostility or retaliation" (Bowen, 1951, 8). His attitude was "to offer my help and support … to meticulously avoid anything … construed as advice or a suggestion or to minimize the problem" (Bowen, 1953, 11). This required a new language with the patients. With fewer expectations to meet, the patient's energy would be diverted from responding/reacting to the relationship to allowing innate responses to emerge. The phrase Bowen coined, "wiggle room," captures what this method provided to the patient.

The most important outcome was the achievement of a state without symptoms, or a "normal child state" (Bowen, n.d.c, 5) that went well beyond what had been considered an adult non-symptomatic schizophrenic state.

> For instance, nurses were active in feeding the patients. He was spoon fed unless he initiated movement to feed himself. His arms and hands were inactive. Arm and hand movement was clumsy and misdirected with poor control over finger and face movement. He used his entire hand, and not his fingers, in putting food in his mouth. This the 'normal child' state was, along with the mental spontaneity, frankness and openness had also a 'normal child' state of very real personal helplessness like a 'normal child'. It was as if he could not do things for himself …. In the 'adult schizophrenic' the personal helplessness was gone but the psychotic symptoms were intense.
>
> *Bowen, n.d.c, 5*

These observations prompted an understanding of the conditions involved in both non-symptomatic and symptomatic states. A goal was set to see through the helplessness and to relate to the "adult" when engaging with the patient. This required listening closely for adult content in the patient's conversation–hearing through the infantile presentations until a thought from and for self emerged. Clinicians working with chronically impaired patients today are familiar with listening carefully, being present and available, not caretaking or offering advice, and not directing the person toward a solution while waiting for the "I" to come out. Staff tried to recognize these adult motivations while being unequivocal about what they could or could not offer in support. By keeping individual choice on both patient's and caretaker's sides of the clinical equation, the environment became a place where differentiation, or an effort to grow up, could occur. Bowen's operating principle, to combine a changed theory with clinical practice in his search for further clues, was the same method he would use later at NIMH.

Bowen engaged a family's emotional life in how he structured the regression. The family was its own control group. The intense interdependence, confirmed by information available from the social worker, was described in the literature as symbiosis. Whereas Freudian practices focused on the therapist-patient relationship, keeping it pristine by avoiding direct family contact and operating on the premise that transference was a replication of a patient/parent relationship, the regression effort explored a replication of family transference through the doctor/nurse/patient interactions. The nurse/mothering figure and clinician/father figure directly replicated the three-person original relationship configuration in the family. Bowen altered the basic two-person Freudian approach to represent a nurturing family but with dedicated efforts to respect the individuality of these adult (offspring) patients.

A boundary was maintained, thereby preventing fusion of the caretaker's self and the patient's self. Individuals remained resources to each other. The psychoanalytic method of attending counter-transference, the clinician's managing of personal reactiveness originating in his or her own sensitivities formed in that parental relationship, was half way to this effort to avoid forming a transference. This valued each individual's innate capacity for taking responsibility for self.

The psychiatrist and nurses' heightened awareness of subtle mothering demands and their removing them made the difference. Regression turned out to be self-limiting, lasting from 24 to 28 weeks, ending spontaneously by the patient seeking more independence. Growth was innate. Bowen learned what conditions prompted this.

Once improvement occurred, the regression program was ended. Then Bowen offered supportive, intensive psychotherapy. This did not involve the doctor and nurse in daily hands-on care of the patient. Once no longer psychotic, the patient could live nearby or board with local families. The patient selected the frequency of set appointments and chose discussion topics. Bowen offered to continue seeing the in- or out-patient for as long as the person found it helpful, while the social worker continued seeing the family. Support was directed at increasing the person's comfort level and awareness of competence for making his or her own decisions. The emphasis was on continuing a relationship. This two-pronged approach, regression therapy, then, following improvement, intensive supportive psychotherapy and ongoing social work meetings with family, continued until Bowen's departure from Menninger.

In Bowen's regression work at Menninger, it was common to have years of therapeutic contact with in- and out-patients before a patient was discharged to the outside world. The regression method was an effective treatment if the psychiatrist devoted his professional life to the treatment of just a few people. The real gains from this effort were (1) recognizing the subtleties and attitudes that intensified or de-intensified the relationship between patient and therapist; (2) observing a basic self and functional self, which gave new ideas on differentiation of self and an emotional system; and (3) the postulation of a relationship while leaving the transference within the family (Bowen, 1956c, 2). Bowen's new theoretical assumptions were worthy of wider application.

Based on my review, by 1951 a different understanding had formed, namely, that rejection/deprivation was the negative side of a strong attachment process. In a symbiosis, the positive side was an overprotective parent. Either parent or offspring showed an "I can't live with you" as the negative side or an "I can't live without you" connection to the other. If the therapist, functioning as the parent, neither encouraged growth nor withdrew from the relationship or, said another way, held steady by being a presence to the other, a new cycle toward progress began. These cycles repeated. When the psychiatrist could stay steady, the person could use the relationship as a home base. Regressions were not as deep, fixed or prolonged. These observed cycles of closeness and distance later became a teaching tool for the staff at NIMH on expected observations of the mother–daughter pairs.

Circumstance and personal interest coalesced in 1948–1951. He was searching for clues about human functioning and the ways in which family relationships fit. The leaders at Menninger thought of psychiatry as a discipline with potential for grand discoveries for humankind. The expectation for staff to advance what they knew about human problems gave Bowen leeway for his in-patient studies. Serendipitously, in 1949, he cofounded the Shawnee Guidance Center in Topeka. It became in 1951, the place where Bowen could further his studies with an out-patient population, thus checking for generalizability, a quality required for any solid theory. This is taken up in Chapter VIII.

Note

1 There is some discrepancy in the archives as to the diagnostic category of people with whom he used regression. "Four schizophrenic patients and four severe alcoholic patients participated. None had improved with long periods of other therapeutic approaches" (Bowen, 1962, A4). Another note says,

> This the beginning of an experimental treatment program used with 8 patients over a period of three years. The patients included 4 schizophrenic patients and two severe alcoholic patients who had not improved with long periods of other therapeutic approaches.
>
> *Bowen, 1962, W4*

The diagnosis of the other two participants is unknown, though from my research, I believe one was a young woman in an incestuous relationship with her father. Exactly who the patients were is an area needing further research.

References*

*Unless otherwise specified, the works of Murray Bowen included in this chapter are from the Murray Bowen Papers. 1951–2004. Located in: Modern Manuscripts Collection, History of Medicine Division, National Library of Medicine, Bethesda, MD. The Accession, Box and Folder information differs for each one and that is included here.

Bowen, M. (n.d.a). [Draft. "II Theoretical background for project"]. (Acc. 2014, Box 6, Folder Chapter I. the research project).

Bowen, M. (n.d.b). [Draft on regression work]. (Acc. 2005-055, Box 3, Folder Developing ideas, working papers & letters).

Bowen, M. (n.d.c). [Draft]. (Acc. 2005-055, Box 3, Folder Working papers).

Bowen, M. (1947). [Diagram of patient's family, January 1947]. Copy in possession of Bowen family, Williamsburg, VA.

Bowen, M. (1948a). [Letter on April 5, 1948]. Copy in possession of the Bowen family, Williamsburg, VA.

Bowen, M. (1948b). [Presentation to District Rotary Club Convention, Topeka, KS]. Copy in possession of the Bowen family, Williamsburg, VA.

Bowen, M. (1948c). [Letter on October 17, 1948]. Copy in possession of the Bowen family, Williamsburg, VA.

Bowen, M. (1948d). [Letters, February 25, March 10, March 28, 1948]. Copies in possession of the Bowen family, Williamsburg, VA.

Bowen, M. (1949). [Menninger, clinical correspondence]. Copy in possession of the Bowen family, Williamsburg, VA.

Bowen, M. (1950a). [Letter on January 1, 1950]. Copy in possession of the Bowen family, Williamsburg, VA.

Bowen, M. (1950b). [Letter on April 8, 1950]. Copy in possession of the Bowen family, Williamsburg, VA.

Bowen, M. (1951). [Presentation: "Direct gratification of the infantile need in schizophrenia", October 1951]. (Acc. 2006-003, Box 5, Folder Papers, schizophrenia notes, etc.).

Bowen, M. (1953). [Handwritten draft. Menninger clinical record]. Copy in possession of the Bowen family, Williamsburg, VA.

Bowen, M. (1956a). [DESIGN I. Concept of treating institution re patient, family and itself]. (Acc. 2006-003, Box 6, Folder Formulation of family study project).

Bowen, M. (1956b). [DESIGN II. Concept of treating institution re patient, family and itself]. (Acc. 2006-003, Box 6, Folder Formulation of family study project).

Bowen, M. (1956c). [Family patterns in families with a schizophrenic family member, combined clinical-staff presentation NIH October 1956]. (Acc. 2006-003, Box 6, Folder Family patterns in families with schizophrenic members-Chestnut Lodge meeting).

Bowen, M. (1957a). [A psychological formulation of schizophrenia]. (Acc. 2006-003, Box 2, Folder A psychological formulation of schizophrenia).

Bowen, M. (1957b). [A psychological hypothesis of schizophrenia]. (Acc. 2006-003, Box 8, Folder A working psychological hypothesis of schizophrenia).

Bowen, M. (c. 1957a). [Therapeutic and theoretical gains from experimental regression programs]. (Acc. 2007-073, Box 3, Folder Working papers from NIMH project re: Schizophrenia).

Bowen, M. (c. 1957b). [Draft. A family concept of schizophrenia]. (Acc. 2007-073, Box 3, Folder Working papers/schizophrenia).

Bowen, M. (1959). [Letter on February 19, 1959]. (Acc. 2006-003, Box 8, Folder Board file—Dr. Bowen).

Bowen, M. (1962). [Draft, Family psychiatry]. (Acc. 2014, Box 6, Folder Chapter I. the research project).

Bowen, M. (1963). [Letter to President of the Washington Psychoanalytic Society on February 10. 1963]. (Acc. 2007-012, Box 4, Folder Misc.).

Bowen, M. (1965). [Paper titled "Theoretical and technical approach to family psychotherapy in office practice", South Florida Psychiatric Society, Miami, Florida, December 13, 1965]. (Acc. 2007-073, Box 3, Folder Working papers).

Bowen, M. (1972). On the differentiation of self. Chapter 21 in *Family therapy in clinical practice*. New York, NY: Jason Aronson.

Bowen, M. (1974). Societal regression as viewed through family systems theory. Chapter 13 in *Family therapy in clinical practice*. New York, NY: Jason Aronson.

Bowen, M. (1978). Introduction. In *Family therapy in clinical practice*. New York, NY: Jason Aronson.

Bowen, M. (1987). [Transcription of presentation at Georgetown family systems symposia on September 12, 1987]. (Acc. 2004-013, Box 3, Folder Symposium—Nov. 7–8, 1987).

Bowen, M. (1995). "A psychological formulation of schizophrenia", by Murray Bowen. *Family Systems Journal, 2*(1), 17–47.

Fromm-Reichmann, F. (1948). Notes on the development of treatment of schizophrenics by psychoanalytic psychotherapy. *Psychiatry, 11*(3), 263–273.

Levy, D. M. (1931). Maternal overprotection and rejection. *Archives of Neurology and Psychiatry, 25*, 886–889.

Nichols, C. R., & Bressler, B. (1958). Anaclitic therapy. *The American Journal of Nursing, 58*(7), 989–992

Raz, M. (2010). Anaclitic therapy in North American psychoanalytic and psychiatric practice in the 1950s-1960s. *Psychoanalysis and History, 12*(1), 55–68.

Rosen, J. (1953). *Direct analysis: Selected papers*. New York, NY: Grune & Stratton.

Ruffalo, M. (2018, January 24). The psychoanalytic tradition in American society: The basics. *Psychiatric Times*. www.psychiatrictimes.com/history-psychiatry/ psychoanalytic-tradition-american-psychiatry-basics.

Schlesinger, H. (2007). "American Imago", *64*(2), 229–240, The Johns Hopkins University Press.

Sullivan, H. S. (1927). The onset of schizophrenia. *American Journal of Psychiatry, 84*, 105–134.

Sullivan, H. S. (1931). Environmental factors in the etiology and course under treatment of schizophrenia. *Medical Journal and Record, 133*, 19–22.

Weisbart, C. E. (2014). Transference. In J. Michie, (Ed.), *Reader's guide to the social sciences* (p. 1682). Routledge. https://doi.org/10.4324/9781315062150.

6

SPECIFIC CASE DISCUSSIONS

In-patient studies, 1949–1951

Bowen's regression explorations of in- and out-patients with two different diagnoses are the core of his theory building in this era. I review two subjects here who were identified in archival materials as important to Bowen's theory formulation. While Bowen was the research observer of the patients' experience in the program, it was their motivation to take part that brought returns not only to them but also to Bowen's larger study of human life.

Others watched the research

Because the regression efforts were in plain view, the patients who were not part of the study could see what the Hospital Physician (Bowen) and the nurse were doing. Bowen's access to every patient on the ward meant he heard their reactions along with those of staff. Not unexpectedly, comments from those peripheral to the regression treatment ranged from mild to harsh. The program generated both wishfulness about the special attention the selected patients received and vigorous criticism of the poor treatment he gave these patients (Bowen, 1951a, 20).

A young man with catatonic symptoms of schizophrenia

There are extensive clinical records of a young man Bowen saw for five and half years, beginning in 1948. First hospitalized at age 18 following his first week away at college, the man fit with what came to be Bowen's understanding of the precipitation of a psychotic break. In a symbiotic relationship, the child becomes vulnerable when the relationship with mother is disrupted and there is no viable replacement for her. For this fellow, the external circumstance of going off to college met that criterion. His physical size and his maturity had not advanced in unison. Evaluation

DOI: 10.4324/9781003027287-7

notes from one analyst suggested lobotomy while other psychiatrists described the young man as a "nuclear schizophrenic" (Bowen, 1956a).[1]

As the daily evaluator of patient progress and as the person in charge of activity therapies on the ward, Bowen had leeway in the services people received. In the first period of hospitalization, 1948–1949, Bowen tried various methods with this patient including a "direct interpretation technique" (Bowen, 1951a, 6) and "a period of using TAT cards to get him to project his conflict." Staff, including Bowen, encouraged the patient to conform to expectations such as improved appearance and participation in activities, which sometimes had a positive and sometimes had a deleterious effect. There was no consistency in the man's response. For example, Bowen's notes describe this patient arriving unkempt in Bowen's office, wiping his runny nose on his sleeve. "I looked as stern and angry as is possible with old pitiful, beat up [name redacted]. I said, 'I'll give you five minutes …'" (Bowen, 1950, 2). Finally, the fellow returned to Bowen's office with a comb and handkerchief. Bowen then combed his hair and wiped his nose while the young man complained he could do this himself. Incidents like these provided materials for Bowen's initiation of alternative methods (2).

The regression program began in its first form in 1949. "He has never had any real love …. So the first management device is 'giving' love unsolicited" (Bowen, 1949a, 5). A nurse and Bowen now replicated a family configuration. As a caretaking mother figure, the nurse ignored expectations of dress or hygiene, even discouraging them, while offering individual companionship. This approach allowed the fellow to regress, which in his case meant urinary and bowel incontinence. He saved his feces, and he became increasingly helpless and dependent, staying in bed, and not bathing (Bowen, 1995, 25). Bowen would see him once or twice a day for about 30 minutes of individual attention. The patient's condition deteriorated to a vegetative state, with increased psychosis and withdrawal. The young man claimed the lack of activities was the problem, but when he gained permission to be more active, he did not hold up and he became symptomatic again. This regression approach produced a negative reaction in the staff. The nurses felt that such a plan was parallel to neglect (Bowen, 1951a, 11). In addition, shifts in the patient's responses corresponded to shifts in the treatment's direction (Bowen, 1957a). It was evident that individual psychotherapy and an early version of regression did not affect positive change. The therapeutic effort was at a stalemate; Bowen's learning was not.

> This program lacked design … the nursing staff was made anxious by regression. To give the program structure and goal, it was conceptualized in a different way. This was for the benefit of the staff and not really for the patient … the goal was to reach a point of regression in which the patient could be bottled fed. Immediately there was great interest in this patient who had previously been neglected ….
>
> *5–6*

Archival materials such as this note offer important insight into the regression program and how it changed over two years. Bowen had invited the mother to come for a visit because he wondered if his lack of ability to directly observe family members skewed his understanding of the patient's prior life experiences. The patient's mother came intending to stay a few days in January 1951 and stayed six weeks (Bowen, 1951a, 9). With the next therapeutic plan, both mother and nurses directed their efforts to facilitate the adult man's regression. Each day she came and took her son on pass, returning each evening (11).

Two crucial observations came from seeing both mother and patient together daily in his office (9–12). For example, the young man was sitting on the floor by Bowen's knees and, from time to time, Bowen would wipe the fellow's runny nose. Bowen asked the mother to change places with him. First, Bowen observed that when the mother tended to her son, the fellow appeared more relaxed than Bowen had ever seen. Then there was a display of affection, with the mother kissing him. She remarked this was very unusual in her family. Bowen also observed the mother's inability to give without inserting a demand into the giving. Mother didn't just wipe his nose; she insisted he blow his nose. Bowen was always looking for explicit illustrations to further explore in his search for the biological basis of human behavior. So these observations brought about a new therapeutic course.

The next plan was to mother the young man with no demands (13). Rather than encourage or discourage adult behavior and activities, the "parent" staff allowed the patient to regress, wherever it would lead. His every ask was responded to positively. In the belief that every individual has ACTIVE inclinations toward growth, the intention of the treatment program was to aid this growth possibility. The human is a biological being, the man responds to the environment, nature built in growth processes, and the environment can help growth. As the patient became more baby-like, directed nurse-mothering could stimulate him to grow up psychologically. The accepted idea at the Menninger Foundation was that a symptom was not a weakness or a pathology (Bowen, 1955a, 15) but was restitutive. Bowen, however, considered it "an evolutionary process" (Bowen, 1957b), the effort of the patient to restore balance within self and advance developmentally. It was a strength. People did their best with their own problems. This crucial piece of thinking brought a subtle change in the caretakers' attitude toward unconditional giving. They self-scrutinized their own responses to remove any possibility of expectations. Being present, relating to the patient, and giving without expectation or positively directing the patient was a small refinement but gave impressive results.

The staff could better accept meeting "infantile needs" with "complete mothering" now that the goal of the patient's regression was growth. This allowed for more giving "to make it possible for him to accept mothering" (Bowen, c. 1957a). This point is significant, as most of the caretaking tasks related to basic primary needs of a baby: feeding, cleaning after soiling, changing urine-soaked clothing, bathing and dressing, but on a full-sized adult. Each nurse and psychiatrist had to work on their own attitudes and willingness to do such tasks without resentment (Bowen, 1951a, 21). For example, this fellow saved feces. Managing their own repulsion, the staff respected the importance of the feces to the young man and disposal was his choice. He eventually disposed of the feces or asked staff to do it. At that point, Bowen suggested saving it in a plastic container (22) and not in drawers or his clothing, and the young man accepted this. The fellow went to the nurse's station and asked for the container, placed the feces there and the nurse cleaned him.

The most intense period of the regression covered a five-month period (Bowen, 1951a, 16). When psychotic, the young man was "sloppy, withdrawn, impersonal, lacking any variable affect, distant, insensitive to the environment, interpretations are meaningless" (Bowen, 1957b). Further, he had an "odd bizarre gait, extreme verbal compliance to any comment and extreme negativism" (Bowen, 1995, 24). Then, unexpectedly, the young man was no longer psychotic. It was in the last effort, where the psychiatrist and nurse assumed responsibilities much as a father and mother assume with a six-month-old infant (24), that Bowen discovered the conditions where the person was symptom free. These conditions were consistent across the patients treated with the regression method.

> The initial regression was very rapid, even to a stage of wetting and soiling himself but he would not go on to this stage of bottle feeding ....The essential features had been to expect absolutely nothing of him and to give just as much constant nursing attention as possible. We found that it was impossible to 'expect nothing' .... To speak was to 'expect' a reply. To talk of a job or school was to 'expect' him to improve .... It was found that a hard tone of voice or the need of a nurse to hurry his spoon feeding would cause him to reject further attention all day .... He reached a state at which he 'regressed below the level of schizophrenia.' When maximum efficiency in the program could be achieved, he would be fairly comfortable in the regression, and at this point HE WAS NO LONGER SCHIZOPHRENIC [caps in original] ... he was a normal infant.
>
> *Bowen, c. 1957b*

He became like a normal three- or four-year-old child (Bowen, 1995, 24), "open, frank, direct, at once engaged with anyone in his vicinity, able to express anger and warmth, responsive to the environment, directly engaging staff, overly sensitive to the environment, awkward, interested in changing himself, interpretations are meaningful" (Bowen, 1957c). Noted, without elaboration, was that even his musculature had characteristics of a young child (Bowen, 1957b).

This childlike state was not steady. It was keenly sensitive to and dependent on the environment. Bowen identified the conditions that prompted the patient to "grow up" into psychosis: An unpredictable environment, a perceived demand from another, a perceived demand within self, forced mothering and lack of clarity between staff and patient on the treatment plan. When the patient became once again psychotic, it sometimes took weeks to return to the childlike state (Bowen, 1957a).

> The goal then became one of making it possible for him to stay in this non-schizophrenic infantile state in the hope he would slowly grow up without having to jump back to becoming an adult schizophrenic.
>
> *Bowen, c. 1957a*

While the designated "motherly" nurses provided hands-on care, Bowen met twice a day with the young man in therapy sessions. During Bowen's regression experiments, he was supervised by Milton Wexler,[2] an American lay psychoanalyst, who

> suggested ... that I sit with the patient and free associate 'by him' by which is meant to free associate in his presence and not ... to him. With such an approach, the patient is not expected to listen, to accept, to reject, or anything else. He just listens and makes a comment if he wants to or says nothing if he wants to. This has turned out to be one of the most important factors.
>
> *Bowen, 1951b, 2*

Following Wexler's guidance, Bowen's progress notes recorded he no longer used interpretations (an educated guess about what the patient unconsciously left out) in the treatment beginning in the spring of 1951 (9). Instead, Bowen was present, available and interested without any suggestion he knew more about the young man than the man himself did. Bowen gave direct responses or metaphors. The therapy effort went toward understanding directly the young man's problems. Silence was acceptable in the clinical visit. The psychiatrist introduced no dynamic material.

> One of the very beneficial things ... is somehow to arrange one's attitude ... so as to respect the individuality of the patient and to approach the patient on an equal man to man level ....The point which seems to be the one outstanding point came from an idea from Doctor Wexler ... of putting myself into the treatment ....This really to me seemed to be the missing link I had been looking for.
>
> 9

This quote captures Bowen's shift to a treatment method that encapsulated being present and separate simultaneously. The link that operationalized the life force is being present as a mature self and relating to the mature part of the other without infantilizing or "mothering" the other. Recognizing its importance, Bowen at once integrated this practice into a "non-mothering" effort with out-patients, wanting to test it in a setting with less impaired people (I discuss this effort in Chapter VIII). Besides Wexler, Bowen attributed this idea to hearing similar remarks in a presentation by an unidentified visiting English analyst (Bowen, 1952a, 2).

On a trip in June 1951, Bowen met with this patient's father at the father's office (Bowen, 1951b, 4). There had been extensive correspondence between them before the meeting. During the meeting, Bowen observed that patterns the father described between himself and his son had also occurred in Bowen's own relationship with the patient. This meeting gave him new understandings of the family emotional process that would not have been possible without direct contact with the father.

> I would like to add that this visit with the father on his home ground was extremely beneficial to me in my understanding of the father and the family.

> After having known and visited with innumerable relatives here in the Clinic setting, I was much impressed by the different kind of picture I had by visiting the family in their home setting ... this is so important that we should take the time and the expense to go to the home setting to visit the family in any severe problem such as this when we go into long term treatment.
>
> 5

After this interview, Bowen wrote that he visited his own relatives and had a personally beneficial experience (5).

Results of using a broader lens

This hospitalized fellow was one of a few whom Bowen tried the non-mothering approach with as an in-patient. In doing so, Bowen took the therapeutic stand that this young man's helplessness was "a REAL feeling" but as the patient's therapist he did not have to regard that as a "reality fact" (Bowen, n.d.a, 20). "We never discuss a subject he does not introduce" (Bowen, 1952a, 2). Along with that, Bowen used the same approach with the parent. Bowen was present for the father through correspondence and a personal visit but made no decisions, leaving those between father and son "... getting out of the middle-man position" (3). There was no censoring of mail between son and father (2). The April 1952 progress notes showed reciprocity in the family. Bowen wrote, "Recently this patient has been showing more and more strength within himself and as he shows more strength, the parents became more agitated on the other end of the line" (Bowen, 1952b, 1). The regression diminished when the refined treatment plan was in place and after six to seven months (in mid-1952) it had ended. The patient accompanied Bowen to the National Basketball playoffs in spring 1952, and in summer the patient accompanied a fellow patient on a trip to Texas for six days, in both cases doing well (2).

By fall 1952, Bowen moved up a generation and was taking up similar issues with the father, who expected the hospital to tell him what to do for his pending visit to his son. Bowen gave no advice, just recognized the father's capacity to think for himself (3). The son could not decide whether to accept his father's visit, and it was only an hour before the father's arrival that the patient agreed to it. This was an example of leaving the family's indecisiveness between them, where it originated. This visit between parent and son provided "an early beginning of communication" between them (3). By late summer in 1953, the notes said, "... his improvement is solid and real. There is really no schizophrenic symptomatology and there hasn't been for a long time" (3).

In the summer of 1953, son and father worked out an agreement for the son to return home to the family for a two-week visit (Bowen, 1953a, 4). After a prolonged visit at Christmas in 1953, the father wrote to the Menninger Foundation that his son's functioning at home with the family was near normal (Name redacted, 1954). At the end of 1953, Bowen told the patient of his plans to leave Topeka within six months and that the patient could transfer to National Institute of Mental Health

(NIMH) and continue a relationship with Bowen if he wanted (Bowen, 1954a, 2). In June, a few days after Bowen left Menninger, he sent a letter to the father reflecting on his relationship with this young man.

> It has had its rewards for me that will be reflected in the life of every person I shall ever touch in the future and for that I am and shall always be grateful …. Changes have been very slow … it could be said that the impossible has changed to the possible …. I think that somehow [your son] will emerge with an adjustment satisfactory to him and with a capacity to use his unusual gifts of sensitivity, compassion and kindness. I hope so. The world can use it.
>
> *Bowen, 1954b*

In the ten months before Bowen left, in June 1954, his approach to the family was consistent.

> I have continued to hold the hospital attitude of expecting nothing and giving as much as possible, and in the psychotherapy I have attempted to hold a very neutral attitude expressing no opinions or values, leaving decisions up to him and taking an active stand against the patient's constant efforts to have someone take over and direct his life.
>
> *Bowen, 1953a, 5*

Bowen expected that this young man with schizophrenia would, at his own request, follow him to NIMH. By May 1954, the patient informed Bowen he would stay in Topeka (Bowen, 1953a, 3). The fellow visited NIMH in August 1954 but went into out-patient treatment in Kansas. Correspondence with this family continued through 1958.

Regression treatment with alcoholism

For both alcoholism and schizophrenia, Bowen planned the same regression program with one exception. Insulin, discovered in 1921 and treating diabetes the next year, was being used to treat psychiatric disorders by 1926 (Wortis, Bowman, & Goldfarb, 1940, 671. Insulin coma or insulin shock therapy was a treatment for schizophrenia. It was used to treat alcoholism by 1940 (Wortis, Bowman, & Goldfarb, 1940, 675–678). Bowen used it with alcoholics in his regression program.

> The post insulin coma state was one in which patients were very receptive to nursing attention. The states seem to vary with the amount of nursing attention available. For instance, the patient on small doses of sedative insulin was as receptive to attention or even more receptive as the post insulin coma patient.
>
> *Bowen, 1958*

Others had found this receptiveness as well. "During waking, and for some time after, patients are in a particularly receptive, 'pliable' frame of mind, and more responsive to common-sense psychotherapy than at any other time" (Pullar-Strecker, 1945, 19).

Bowen reasoned that patients coming out of insulin coma would gain more benefit from the milieu:

> that if the problem was a deprivation of love, and if in the twilight of insulin the patient could accept the therapeutic attention, then the treatment program could be shortened by the creation of some kind of situation in which it was possible for the patient to accept love and attention and for the environment to give as much as the patient could accept.
>
> *Bowen, 1957a, 3*

Case with alcoholism

The first person with alcoholism receiving this treatment program had a 12-year history of alcoholism. He had been in psychoanalysis five times a week for over three years, received in-patient care at Menninger, and was viewed by clinical staff as a "treatment failure" (Bowen, c. 1957a). As ward physician, Bowen knew him from his first admission to the hospital, in August 1948. He was a 30-year-old married man from a Southeastern coastal state, the eighth of nine children and the youngest of six brothers, all college graduates (Bowen, 1948). He worked in the family business, an equal partner with his brothers. The mother died when he was in his third year at college. "'The reason for living' was gone from his life" (Bowen, 1953b, 3). "Things were never the same after that" (Bowen, 1953c, 2). The patient began drinking soon after, while attending a prestigious East Coast medical school from which he dropped out after an illness. This man completed flight training for the Navy and arrived intoxicated on the day he got his wings (Bowen, n.d.b, 3). During one of his drunken times, he married a divorced woman he knew only briefly, wanting "someone to take care of him during his drunks" (Bowen, 1949b, 3). His family did not readily accept the wife. The couple and their two young children became isolated when he was working long hours. The patient had a prolonged period of hospitalization, from 1948 to 1951, with Bowen as his hospital physician, and with a separate analyst. Finally, the analyst told him if he drank again, the analyst would not see him. The predictable outcome was that the patient drank again, and the analyst refused further contact (Bowen, 1990, 1). While psychoanalysis had given this fellow superior intellectual insight, it had not interrupted the drinking (Bowen, c. 1953, 6). Bowen had multiple contacts with the wife during this time. She had moved from another state to Topeka to visit her husband (Menninger clinical record, 1949a, 1). The social worker saw her at various times, but the wife discontinued contact when she felt "it implied she was sick herself" (7) and returned when her distress level became unmanageable.

A staff conference in November 1949 on this patient clearly illuminates psychoanalytic thinking and treatment in that time and place. The social worker noted eloquent discussions in meetings with the wife about her struggles (1–3). The notes showed the dependency between the two spouses, yet ignored the influence of family and deep connections between family members. The theory of the individual as an autonomous unit with a negative mother complex dominated the notes of the ward psychiatrist and social work and nursing staff. "The indulgent mother is the principle [sic] (traumatic) person in an alcoholic's life" (Bowen, 1949b, 9). The nurse's notes discussed psychoanalytic principles in implementing milieu therapy and ways of giving attention without being caught in a fusion with the patient (Menninger clinical record, 1949b, 1). The staff was close to viewing the family as a unit, but the prevailing way of thinking obscured the concept until January 1951, as a result of Bowen's observations of families and his therapeutic efforts.

Bowen remained the ward physician, and individual psychotherapy for the patient with alcoholism continued under another psychiatrist. After repeated elopements, delirium tremors (DTs) and at least one serious suicide attempt, the patient ended treatment in August 1951 (Bowen, 1985, 1). The patient did not want psychoanalysis as it "offered him nothing for the future" (Bowen, 1955b). A month later, this man relapsed and Bowen agreed to see him clinically, using regression (Bowen, 1990, 2). Their contact before this, while extensive, was not "clinical." In correspondence, Bowen noted that his efforts from 1949 to 1951 explore what he called "research problem[s] with schizophrenia." By beginning a regression program for patients with alcoholism, Bowen added another diagnosis even though his work schedule was full to overflowing up to "54 hours a week," not counting the paperwork (Bowen, 1951b, 1).

Bowen began regression treatment in September 1951. The goal was to replace the dependent relationship with his "dead mother, with a relationship with me" (Bowen, 1990, 2). The patient asked for both insulin and paraldehyde.[3] Having learned from the patient's dismissal of the previous analysts, Bowen avoided using "no." In the notes, he connected using "no" to this fellow's own father's inability to say either yes or no and to the mother's indulgence (Bowen, 1949b, 2). Describing the technical way of doing this, Bowen wrote, "You don't say 'No'—you just mean it without saying it" (Bowen, 1990, 6). "I never told him a direct 'no' but never gave him any reason to believe that I would even consider 'Yes'" (6). Bowen observed the gentleman needed mothering attention as he emerged from the insulin effects. Along with mothering attention provided by Bowen and the assigned nurse was the effort to "wean him from alcohol through the use of paraldehyde" (Bowen, 1953c, 4). Requests from the patient for paraldehyde became part of the therapeutic discussions and were always left to his initiative. Interestingly, in several places in these notes, Bowen related his own concerns about this man's use of paraldehyde. He received support from "other members of the section [who] gave me the courage to stick to the ..." plan (Bowen, 1990, 7). Bowen saw the patient twice a day, an hour in the morning and an hour in the afternoon (2). He had a deep regression by December 1951. Bowen referenced this fellow in the draft of his 1953 research proposal on alcoholism:

> He lived out with the nurse-mother, the doctor-father, and the other patients–siblings what would appear to be an exact replica of the early infancy relationships. He came through this firmly 'imprinted' to the therapist from which we worked psychotherapeutically to a successful resolution of the transference.
>
> *Bowen, c. 1953, 6*

The notes on this fellow showed cycles of closeness to and rejection of both the staff and Bowen. The notes also showed the patient's sensitivity to shift changes and staff replacements, environmental conditions that could prompt an interruption in the regression back to a symptomatic state. Bowen described another observation with this man as "somatic protest" (Bowen, c. 1952, 9). This meant physical symptoms occurred, but lab results were negative. In later writings, Bowen observed sequences in the patient's progression: "shift from whiskey to paraldehyde ... shift from paraldehyde to coffee ..." (Bowen, 1990, 4). He asked, "Can people shift addiction from one agent to another—I know the patient can do it—hard tho it is, but doubt ability for therapist to suggest it" (Bowen, c. 1952, 4). By February 1952, the patient bounced out of it. He was functioning well enough to move to a hotel in town and become a day patient the next day (Bowen, 1953b, 15). By 1953, Bowen wrote,

> During his course of insulin treatment, he developed a transference relationship which was about as intense and positive as transference can be. I use the term transference because it was and is transference in the true sense of the world .... He was constantly trying to set up the situation which would cause me to reject him .... I tried all ways to maintain what I referred to as a 'neutral' attitude .... I tried very hard not to do anything which would go in the direction of 'holding him' in the transference, and nothing that would permit him to 'bolt out of it'.
>
> *Bowen, 1953d*

This example of Bowen managing his countertransference describes his newly implemented efforts to avoid transference. He recognized the emotional demands of the other, keeping emotional contact without withdrawing or providing solutions. As with schizophrenics, the plan used mothering nurses to meet every need for attention or services as much as possible, which were offered as long as the person found it useful (Bowen, n.d.c, 5). Bowen wrote of the method's effectiveness:

> Slowly, the terror of helplessness subsided until there were short periods of relative comfort in which there seemed to be enacted before our eyes the triangular drama of the relationship between nurse-doctor-patient which

seemed to me to be a reasonable facsimile of the original mother-father-infant relationship.

Bowen, 1954c

The observation of the "triangular drama" was not unique to people with schizophrenia or alcoholism, and Bowen went so far as to say it was in everyone "to some degree" (Bowen, 1954c).

Bowen considered the "terror of being helpless," the fear of losing the other, and the terror of aloneness as "equivalent to anxiety" (APA Meeting, 1959), a threat to one's survival. His definition of anxiety came from his participation in the intensity of the symbiosis. As he listened to conversation after conversation with psychotic or chronically impaired people, he distinguished the actual threats these people faced from the threats originating in their minds, or imagined threats. Much of the upset they experienced was around imagined threats. It was difficult for the person to know the difference, but for Bowen, the observer, it was possible. The anxiety about helplessness and aloneness was a response to a perceived or imagined threat, and the response became a rumination mantra that interfered with resolving the helplessness.

This man "never got drunk again" after taking part in the regression program (Bowen, 1985, 1). He had a "regressive fallback" with a brief hospitalization in July 1953 (Bowen, 1954d, 1), and he continued his relationship with Bowen when he transferred to NIMH, in the small in-patient alcohol program Bowen set up. There, his wife was actively involved in that treatment. The clinical relationship lasted from 1948 to 1956 (Bowen, 1965). There is a video of a 1970s interview with this man where he discussed sequences in his addiction: to alcohol, to paraldehyde with attachment to the motherly aides, then to Bowen and finally attachment to his wife. Bowen maintained a relationship with the man and had contact with the extended family at least through 1984.

Similarities between the two diagnostic groups

There were some general similarities between the schizophrenic and alcoholic patients. Both sets of patients regressed quickly, some to the point of incontinence (Bowen, c. 1959, 5). Over time, the depth of the regressions eased; the relationships were more fluid and intense dependency was not as prolonged. The eight patients involved in the regression treatment developed a potent, positive transference or close, concentrated relationships with the therapist, having characteristics, found in the literature at the time, of symbiosis. Attempts to resolve the transference gave insight into the relationships. They were long term and difficult to resolve (Bowen, 1956a, 2). Yet they allowed for a calm discussion of disturbing conflicts at a deep, intense level. This outcome did not support assumptions about rejecting mothers or repressed unconscious hostility in the patient and was replaced by this understanding of a symbiotic relationship.

These relationships could continue week after week at this level. The capacity to sustain this was unusual in other close therapeutic relationships (Bowen, c. 1959, 5).

When pressed to express therapeutic anger, the people did. If there was no pressure, people were not really angry. In Bowen's observations, they were mainly miserable (Bowen, 1962a). When the environment interfered with the close relationship and they attempted to go it alone, psychotic symptoms returned. This could last a few hours or a few days until they were comfortable again in the symbiotic relationship (Bowen, 1957a). Patients spontaneously ended the regressions after 24–28 weeks (Bowen, 1956a, 2), if there was no inadvertent forcing of attention or interruption in the provision of attention. Growth was innate. An initiative to "wean" from the program came from the patients themselves (Bowen, c. 1959, 6). Improvement occurred. Psychotic symptoms disappeared. Severity of drinking problems lessened. The people were "essentially normal infants" in the regressed phase and, given the right conditions, they moved to more maturity within themselves.

There was evidence that the therapeutic relationship replicated the original parental relationship in how the person functioned in relation to the doctor and nurse. The participants who were emerging relatively free of psychotic symptoms had intense symbiotic attachments to the therapist (Bowen, 1956b). Bowen checked for historical evidence in the social workers' information of the same relationship characteristics between the parent and the patient. And when Bowen had direct contact with a parent on their home ground, these relationship characteristics were also present.

Bowen's direct participation and deep involvement, combined with his ability to observe, keep an open mind and allow facts to emerge, had the potential to bring more solid concepts and better outcomes. He had one foot in as a participant in the regression and one foot out as an objective observer of the interactions between himself and the patient. Bowen's intention with his methods was to help people. This version of regression did not depend on theoretical interpretation—just on watching and describing.

> I believed that this phenomenon, schizophrenia, was the product of maternal deprivation. The treatment problem seemed to be one of getting by the rejection barrier with sufficient love to 'heal' the injured ego. Could the whole long treatment process be speeded if there was a way to 'get to' the patient with concentrated loving attention? The program was geared to continue indefinitely or until the patient showed signs of growing up on his own …. All the patients went through 3 stages. The first stage was one of alternating acceptance and rejection of attention and an increase in psychotic symptoms. The average stage was about 8 to 10 weeks. Then came a stage of more or less complete acceptance of the passivity and attention. In this period, the psychotic symptoms suddenly disappeared, and the patients became like comfortable contented children. Their open, naïve frankness was like a child. Previously non-communicative patients could relate openly and freely about personal subjects. The last stage was the renunciation of attention and development of self-growth. First, they begged that attention be stopped, and they

be forced to activity. Then they were able to renounce this on their own. They would try to jump from regression to adult behavior. Symptoms would return immediately. Then they learned if they paced their move to adult behavior, symptoms would stay away …. The experience began to provide more and more material that could not be explained by rejection theory. Especially in the stage of attempted resolution of the relationship, when the patient was as comfortably verbal as in the stage of acceptance of attention, there were repeated expressions to indicate the relationship to the mother was more of a basic over-attachment than a basic rejection.

Bowen, c. 1959

Differences in results

The alcoholic was more concise than the schizophrenic and more sure in demands, repudiation of attention and denial of a wish for attention. In subsequent years, Bowen referred to this as a clean symbiosis. The schizophrenic's symbiosis was dirty or muddled. Prior to the regression treatment, two patients with alcoholism had severe drinking problems over many years, including lengthy histories of DTs. People with chronic alcoholism could ask for and did receive insulin (Bowen, c. 1959, 5). Two of the patients with alcoholism used insulin regression to help them accept the mothering. Two alcoholics became social drinkers in the two-to-three-year phase where the therapist was "home base," including the individual discussed earlier in this chapter.

A mother of a third alcoholic reactively removed him from treatment when she found the intense positive relationship with the therapist intolerable. She transferred him to a state hospital, and he continued to be a problem drinker. Yet after ten years, he never had DTs again (Bowen, 1962b). I do not have information on the fourth patient.

In the four patients who had serious schizophrenic impairment, none completely resolved the symbiotic relationship. One married and lived a productive life, another had a good adjustment in a protected environment, and a third transferred to a state institution with two brief hospitalizations (Bowen, 1957a). The fourth achieved a partial resolution. "All seemed to have benefitted by the treatment experience …. The regression program was most helpful to the therapist … it revealed the therapist's subtle ways of mothering the patients that had not been recognized in other psychotherapy" (Bowen, 1957d, 11).

Later, after his NIMH research, Bowen no longer considered schizophrenia to be resolvable within the nuclear family, writing of his continued study on the subject:

It suggested that the symptom complex known as schizophrenia is a manifestation of a faulty level of 'differentiation of self' in the human environment, rather than a disorder confined to schizophrenia alone.

Kerr & Bowen, 1988, 364; italics in the original

Foreshadowing

The basic method of Bowen family system therapy was now in place: an interested, knowledgeable and detached observer provides a forum for the individual's efforts with their actual family members to be more mature. The change effort takes place in the family. While at Menninger, Bowen spoke to colleagues about staying out of the transference and reported, "disbelief from other professionals that there was such a possibility who then reframed it to 'You mean you handle it well'" (Bowen, c. 1959).

Refinement of his ideas throughout his lifetime is a distinctive attribute of Bowen's writings. By 1975, he wrote,

> In years past I used to get on the relationship 'between' people. It took me half a lifetime to discover relationships automatically change if the same person (therapist) can stay unbiased, and help each 'self' become a better defined self.
> *Bowen, 1975, 1*

Bowen could use the intensity of infantile helplessness as a measure to conceptualize problems of a lesser degree on a continuum. Directing treatment to the patient's adult state meant recognizing and responding to adult expressions and engaging neutrally or even humorously with the patient's infantile-self remarks. The clinician functioned as a catalyst that activated inner growth in the other. Wondering how this might apply to people with less severe symptoms, Bowen set up a project with out-patients to expand on this idea.

Notes

1 According to the American Psychological Association, a nuclear schizophrenic is a type of schizophrenia in which the defining features, including social inadequacy and withdrawal, blunted affect, and feelings of depersonalization and derealization .... It is of early, insidious onset and is associated with a degenerative, irreversible course and poor prognosis (APA Dictionary of Psychology, 2020).
2 Well known for his success with people with schizophrenia while at Menninger, Wexler later founded the Hereditary Disease Foundation and in 1993 identified the gene for Huntingdon's disease. This discovery pointed to the possibility, later achieved, of mapping the human genome.
3 Paraldehyde is a strong sedative used historically for childbirth pain, alcohol withdrawal and to stop seizures.

References*

*Unless otherwise specified, the works of Murray Bowen included in this chapter are from the Murray Bowen Papers. 1951–2004. Located in: Modern Manuscripts Collection, History of Medicine Division, National Library of Medicine, Bethesda, MD. The Accession, Box and Folder information differs for each one and that is included here.

APA Dictionary of Psychology. (2020). *Nuclear schizophrenia*. Retrieved March 24, 2020. https://dictionary.apa.org/nuclear-schizophrenia.

APA Meeting. (1959). [Impressions from the 115th annual meeting of the American Psychiatric Association, April 27–May 1, 1959]. (Acc. 2006-003, Box 1, Folder Documenta Geigy).

Bowen, M. (n.d.a). [Draft]. (Acc. 2007-073, Box 3, Folder Working papers from NIMH project re: Schizophrenia).

Bowen, M. (n.d.b). [Handwritten draft]. (Acc. 2007-073, Box 3, Folder Working papers).

Bowen, M. (n.d.c). [Draft. Clinical experiences leading to psychological formulation of schizophrenia]. (Acc. 2007-073, Box 3 Folder Working papers from NIMH Proj. re: Schiz).

Bowen, M. (1948). [Menninger clinical record. Case abstract September 2, 1948]. Copy in possession of the Bowen family.

Bowen, M. (1949a). [Menninger clinical record, Physicians' order sheets, name redacted, December 22, 1948]. Copy in possession of Bowen family, Williamsburg, VA.

Bowen, M. (1949b). [Staff conference on the hospital management of alcoholics on November 15, 1949]. Copy in possession of Bowen family, Williamsburg, VA.

Bowen, M. (1950). [Letter on April 4, 1950]. Copy in possession of Bowen family, Williamsburg, VA.

Bowen, M. (1951a). [Draft. Direct gratification of the infantile need in schizophrenia]. (Acc. 2006-003, Box 5, Folder Papers, schizophrenia notes, etc.).

Bowen, M. (1951b). [Clinical record progress notes, name redacted, August 31, 1951]. (Acc. 2006-003, Box 5, Folder Papers, schizophrenia notes, etc.).

Bowen, M. (c. 1952). [Menninger clinical record alcohol patient]. Copy in possession of Bowen family, Williamsburg, VA.

Bowen, M. (1952a). [Clinical record progress, name redacted, September 15, 1952]. (Acc. 2006-003, Box 5, Folder Papers, schizophrenia notes, etc.).

Bowen, M. (1952b). [Clinical record, name redacted, progress notes on regression patient, April 11, 1952]. Copy in possession of Bowen family, Williamsburg, VA.

Bowen, M. (c. 1953). [Draft. Proposed program for psychiatric research and study]. (Acc. 2007-073, Box 3, Folder Bowen-NIMH-Working papers incl. proposal for proj.).

Bowen, M. (1953a). [Clinical record progress notes, name redacted, August 10, 1953]. (Acc. 2006-003, Box 5, Papers, schizophrenia notes, etc.).

Bowen, M. (1953b). [Draft. Observations on the character structure of the alcohol addict, May 20, 1953]. Copy in possession of Bowen family, Williamsburg, VA.

Bowen, M. (1953c). [Discharge summary, Menninger clinical records. December 1953]. Copy in possession of the Bowen family, Williamsburg, VA.

Bowen, M. (1953d). [Course in the institution and summary of treatment, Menninger clinical records. December 16, 1953]. Copy in possession of the Bowen family, Williamsburg, VA.

Bowen, M. (1954a). [Clinical record progress notes, name redacted, June 1, 1954]. (Acc. 2006-003, Box 5, Folder Papers, schizophrenia notes, etc.).

Bowen, M. (1954b). [Letter to father of a regression patient on June 5, 1954]. Copy in possession of Bowen family, Williamsburg, VA.

Bowen, M. (1954c). [Project paper, fall 1954]. (Acc. 2006-003, Box 5, Folder Project paper).

Bowen, M. (1954d). [Clinical note, May 24, 1954]. (Acc. 2007-073, Box 3, Folder Papers, schizophrenia notes, etc.).

Bowen, M. (1955a). [Presentation on July 7, 1955: Practical frontiers in mental health]. (Acc. 2006-003, Box 5, Folder Mtg notes October 1954).

Bowen, M. (1955b). [Doctor's progress note, admission note]. (Acc. 2006-003, Box 3, Folder Name redacted).

Bowen, M. (1956a). [The family and schizophrenia]. (Acc. 2007-073, Box 3, Folder Bowen-NIMH-working papers incl. proposal for project).

Bowen, M. (1956b). [Presentation to a combined clinical staff meeting at NIMH. 1956]. (Acc. 2006-003, Box 7, Folder Family patterns in families with a schizophrenic family member).

Bowen, M. (c. 1957a). [Paper. Formulation of the family study project]. (Acc. 2006-003, Box 6, Folder Formulation of the Family Study Project).

Bowen, M. (c. 1957b). [Untitled paper. Draft begins "Two years' experience…"]. (Acc. 2006-003, Box 6, Folder Formulation of the Family Study Project).

Bowen, M. (1957a). [Paper. A psychological formulation of schizophrenia]. (Acc. 2006-003, Box 6, Acc. 2006-003, Box 2, Folder A psychological formulation of schizophrenia).

Bowen, M. (1957b). [Draft, "This I believe about schizophrenia]. (Acc. 2006-003, Box 8, Folder A psychological concept of schizophrenia).

Bowen, M. (1957c). [Paper. A family concept of schizophrenia]. (Acc. 2007-073, Box 3, Folder Isolated working papers re: Psychosis, a family concept of schizophrenia by M. Bowen).

Bowen, M. (1957d). [A psychological concept of schizophrenia]. (Acc. 2006-003, Box 8, Folder A psychological concept of schizophrenia).

Bowen, M. (1958). [Letter to George Daniels July 10, 1958]. (Acc. 2006-003, Box 6, Folder 3-E project reprint distribution).

Bowen, M. (c. 1959). [Draft. Family psychiatry] (Acc. 2014-034, Box 6, Folder Chapter I the research project).

Bowen, M. (1962a). [Part I theoretical background]. (Acc. 2014-034, Box 6, Folder Chapter II theoretical background).

Bowen, M. (1962b). [Chapter 1 the research project]. (Acc. 2014-034, Box 6, Folder Chapter I the research project).

Bowen, M. (1965). [Letter, November 20, 1965]. (Acc. 2003-026, Box 5, Folder name redacted).

Bowen, M. (1975). [Letter on March 15, 1975]. (Acc. 2003-026, Box 5, Folder Name redacted).

Bowen, M. (1985). [Letter to the social worker on Bowen's team at Menninger on July 4, 1985]. (Acc. 2006-003, Box 2, Folder Correspondence materials).

Bowen, M. (1990). [Presentation notes for Conference: Addictions and family Systems, Green Bay, Wisconsin, April 20, 1990]. (Acc. 2007-012, Box 2, Folder Working papers).

Bowen, M. (1995). A psychological formulation of schizophrenia. *Family Systems, 1995*(2), 17–47.

Kerr, M., & Bowen, M. (1988). *Family evaluation: An approach based on Bowen theory.* New York: W. W. Norton.

Menninger clinical record. (1949a). [Social work notes, November 15, 1949]. Copy in possession of the Bowen family.

Menninger clinical record. (1949b). [Nursing role in hospital management of the alcohol addict, November 15, 1949]. Copy in possession of the Bowen family.

Name redacted. (1954). [Menninger clinical record. Letter from the father of a patient to Jerome B. Katz, Chief of Section, Menninger Foundation on January 4, 1954]. (Acc. 2007-073, Box 1, Folder Patients Bethesda 1954-5).

Pullar-Strecker, H. (1945). The use of insulin in the treatment of alcoholism and alcoholic addiction. *The British Journal of Inebriety, 43*(1), 14–27.

Wortis, J., Bowman, K. M., Goldfarb, W. (1940). The use of insulin in the treatment of alcoholism. *Medical Clinics of North America*, 24(3). https://ur.booksc.me/book/54547917/b81071

7

SYMBIOTIC RELATIONSHIPS

As Bowen was beginning his own explorations of the depriving or rejecting mother assumption, mother–child symbiotic relationships were also of interest to others. In 1949, Therese Benedek, a psychoanalyst from Hungary who came to the United States in the 1930s, and one of Bowen's sources in his 1978 book, discussed her theoretical perspective of the family as an "organism" that included characteristics of the mother–child symbiosis. She wrote that a child had a unique triangular relationship with each parent and that these "triangles" created the family's emotional environment.

> For, in the child, the mother and then the father, unconsciously relive a specific part of their individual personalities as well as their relationship to each other …. Since each of the parents will project different expectations on to the child, each may experience different variations of his own personality problems with each new child … each triangle unit is dynamically influenced by the other units within the family.
>
> *Benedek, 1949, 215–216*

She included the father and siblings as important in the nuclear family process. Bowen noted discussions of symbiosis in the literature.

> Benedek (1949) had discussed the theoretical aspects of the mother-child symbiosis. Mahler (1952) had discussed clinical implications in her work with autistic and symbiotic children. Hill (1955), Lidz and Lidz (1952) and Reichard and Tillman (1950) had considered symbiosis as it applies to the adult schizophrenic patient.
>
> *Bowen, 1960a, 69*

DOI: 10.4324/9781003027287-8

In the late 1940s, Bowen intended to come to his own conclusion. What would observation of mothers and impaired offspring tell him?

Writing about the in-patient study at Menninger, Bowen noted the qualities of a symbiosis in the relationships between the patient, himself as the father and the nurse as mother:

> By 'symbiotic' is meant the intense emotional attachment between mother and adult child .... For all the questions about the accuracy of the term and comparisons with the strict use of the term in biology, I had in mind the popular use of the term as used in psychiatric practice since the 1930s. The intense rejection of the mother by the patient, or the patient by the mother, or the rejection of each by the other was seen as the negative phase of the intense positive attachment that is more characteristic of the alcoholic patient and his mother.
>
> *Bowen, 1962*

Bowen allowed for the possibility that either parent—the mother or the father—could be the primary participant with the offspring. The mother's pregnancy and early caretaking duties made her the more likely one for this most intense relationship. In Bowen's regression experiment, he said something surprising.

> When the 'treatment parents' had replaced the real parents in psychological value, the male therapist was in the position of the mother, as had been described as mother-child symbiosis in the literature. It would appear that the position of 'mother' had to do more with the function than the sex of the parent figure.
>
> *Bowen, n.d.a, 13*

This insight adds another dimension to Bowen's choice to admit mother and daughter pairs at National Institute of Mental Health (NIMH). Bowen allowed for the possibility that either parent—the mother or the father—could be the primary participant with the offspring. My research indicates that one of his regression families involved a father functioning as mother.

In his experimentation he found "... it was not possible to have a therapeutically moving relationship with the patient, until I was in the symbiosis with the patient, and then the road out was so long and difficult and unpredictable" (Bowen, 1959). By taking part in a symbiotic relationship while attempting to research it, Bowen gained firsthand information on the innate life forces of differentiation (individuality) and togetherness. He identified the conditions that triggered a return of symptoms, suggesting the self played a part in the pull from each of the life forces. This motivated Bowen to leave the symbiosis where it originated and to put his richly resourced therapeutic effort in the biological family.

It was the observations at NIMH that shifted his understanding of the term symbiosis from a psychoanalytic to a biological basis. "The concept of symbiosis, originally from psychiatry, would have been discarded except for its use in biology where the word has a specific meaning" (Bowen, 1976, 354). In the human, a symbiotic relationship was a natural developmental process from birth through early years. A child could not thrive without close interaction with a parent or other primary caretaker. The underlying assumption of maternal deprivation was not born out by the research, whereas the over-attachment observed in Bowen's research participants supported the biological concept of symbiosis.

A biological mother did not have the awareness that the clinician "mother" had. Because of Bowen's ability to remain objective while deeply interacting with patient and family, he understood how the biological mother differed from him as the mother figure. The biological mother was an active participant in the mother–child relationship and apparently depended on the symbiosis, as it had not resolved with ordinary maturational growth. This suggested that the mother began the symbiosis, but both mother and offspring sustained it.

While life sustaining in infancy, prolonging a symbiotic relationship interrupted the developmental phase toward independence. Bowen understood schizophrenia or alcoholism to be an adaptation, a person's way to move out of the symbiosis. Without the symptoms of schizophrenia or alcoholism, the impaired adult was a functional child in an adult's body. With the symptoms, the adult was psychologically impaired, but the symbiosis was manageable to a degree (Bowen, 1957a).

Regression therapy showed that people with schizophrenia and alcoholism, contrary to what was then understood, were capable of a transference. The variation in their functional states depended on the environment and on the relationship alignment. Bowen considered them an extension of normal. Bowen considered the possibility that these patients' relationship was an exaggeration of those in any family. He continued to believe that the human was a part of all life and that evolution acted upon all living things.

After years of studying the phenomena of symbiosis, Bowen addressed the difficulty of defining it.

> No one really knows what constitutes this thing we call symbiosis. No one has ever described it satisfactorily except a Tennessee Williams or a Hans Kafka. To be able to live with it is an achievement. To understand it scientifically is a goal.
>
> *Bowen, 1956*

Bowen used the terms symbiosis and symbiotic to represent the intense interdependence achieved between himself and the patient in his regression explorations. His usage is consistent with the understanding at the time in psychiatry—as a developmental lag (Mahler, Pine, & Bergman, 1975, 182). It was not "in" the individual; rather, it was an exaggeration of the deep interdependency happening in many relationships (Nichols, 2008, 126) and was necessary for growth in early developmental stages. "In this context symbiosis has the same meaning psychologically as biological growth on a physical level ... symbiosis is an interdependent relationship mutually advantageous and necessary to both organisms" (Bowen, 1957b).

Later observations at NIMH enhanced Bowen's understanding, and the definition of symbiosis extended from living and being together to "being sick for the other to be well" (Bowen, c. 1956). This gave substance to his much earlier hunch that the human is a biological being deeply related to other living things; the model is an example of how Bowen integrated ideas from biology into his theory. Bowen documented the change in his thinking.

> I originally used the term 'symbiosis' exactly as it had been used in psychiatry to describe the intense mother-child dependence. I considered dropping the term, especially after a nationally known researcher referred to it as a parasitic relationship .... (At NIMH) we researched the literature and found that biology has defined over thirty separate stages between parasitism and symbiosis .... I have used the term 'symbiotic' exactly as defined by biology, and not as it has come to be used in psychiatry.
>
> *Bowen, n.d.b*

Regression, symbiosis and in-patient observations

In Bowen's regression studies, there were no observable or reported symptoms and no patient complaints when there was intense closeness between the patient and Bowen. Bowen had enough information to know this state was very like the relationship between the mother and patient, except that Bowen was also observing, trying to be objective, and taking notes of their interactions. This clinical experience offered the clue that in the deep closeness of a symbiosis between parent and child, symptoms have no purpose (Bowen, 1957c), although this was not sustainable as relationships respond to constant environmental changes. For the human, a pure, continuously symbiotic state without symptoms is unachievable beyond infancy, when dependency is a fact of survival.

Comments made to him by the Menninger staff gave Bowen other insights into the patient-doctor symbiotic relationship. Remember, this effort took place on a ward and was visible to others. When the staff inadvertently called Bowen by the patient's name or called the patient by Bowen's name, it was a tip-off to the visibility of the positive, intense oneness between the patient and Bowen. Another example was when the staff started speaking of either Bowen or patient and included a remark about the other. The inability to keep doctor and patient separate in their discussions was obvious. They detected the symbiosis with their senses (Bowen, 1951).

Human needs for togetherness and separateness were seen at their extremes in the intense interdependence in the symbiosis. Psychiatrists saw the balancing of these two life forces as a gradual separation and individuation within an individual from infancy to adulthood (Mahler et al., 1975, 4). Bowen understood these forces as extending beyond early human development. They were a characteristic of living things at an emotionally instinctual level. Clarence Oberndorf, who wrote *A History of Psychoanalysis in America*, commented that "… the antagonism between individual and communal interest is present in all animal life and is unavoidable among humans" (Oberndorf, 1953, 192). These forces serve as the base of the human's natural dependencies and effort to differentiate a separate self. A commonality of life forces among living things fit with how Bowen was making sense of human life.

> The average person interprets symbiosis as one form of life preying on another form of life. Symbiosis on the biological level means one form of life which 'facilitates' another form of life. For instance, some of the big fish, whales, are examples. And parasites live on the teeth of the big animals to keep the teeth clean. If you take away the parasites to keep the teeth clean the animal dies. These are forms that are necessary for each other. You can apply that to the human, you see, there are people that are necessary for the life and the livelihood of the one next to him. If the one next to him doesn't function, they die. Anyway, that's a complex thing. A symbiosis is where one form of life is essential to the other. Where one cannot exist without the other. OK, you say one person is going to try to improve his level of functioning. He does that by working on himself, but he damn well better be interested in that other guy. If he's not, he will die.
>
> *Bowen, n.*<u>*d.c*</u>

A symbiotic relationship requires interdependence. In the regression patients, when there was action on their part toward independence, symptoms of schizophrenia or alcoholism returned. But in the parent/child symbiosis observed in the first year at NIMH, when one family member improved, the other had difficulties. Bowen saw this as reciprocity in functioning. When a family member's ability to function in daily life declined, study efforts went to identifying what interfered in the relationship and what preceded the decline. Was there withdrawal from the other or a move toward more independence? These were the conditions Bowen observed in the earlier regression studies. At NIMH, he was seeing the same thing with the actual parent and child. Looking back, this pointed to a systems phenomenon. But symbiosis, as described in the literature at the time, recognized a fragment of the family and extended individual theory (Bowen, 1965, 119), but the prevailing view left unnoticed the role of the family, and it delayed Bowen's progress in understanding the family as a unit or system. This is one example where his established method for approaching science interfered with progress in his thinking. While at Menninger, he was close to a theoretical assumption of the family as an emotional unit but did not yet see beyond the literature or trust his own observations. Or perhaps he just did not have enough data yet to establish this then. "Nobel laureate Albert Szent-Györgyi described this phenomenon as seeing what everybody else has seen and thinking what nobody else has thought" (Szent-Gyorgyi quoted in Grinnell, 2009, 30).

Symptoms, an alert for adaptation

Certain conditions found in the patient's interactions with the environment contributed to the presence or absence of symptoms. Bowen's 24-hour reports from the nurses made it possible to track changes from base state to functional state. Both depended on the relationship and on elements in the environment. If the other in the relationship didn't interfere with a growth move or the growing person could tolerate their interfering efforts, independence could result, at least to a slight degree. Bowen could observe these two states in seriously affected patients and in people with less serious problems, those diagnosed as neurotic.

By the mid-1950s, Bowen conceived of self as all that constitutes any individual. This would include all physical components and the intellectual, feeling and emotional systems and how they have been shaped by a family's history and the individual's present existence.

> For the purpose of this theory, self is considered the total of the intellectually determined beliefs, principles and convictions that govern decisions and one's life course. I find it profitable to think of 2 levels of 'self' one a hard core level of 'solid self' determined by inner conviction and which is not easily changed. The other is a poorly defined level of pseudo self made up of a variety of philosophies, good sounding principles learned from others and adopted more to fit the feeling system than logic, and that can be easily modified by persuasion.
>
> *No author, c. 1957*

In the regression patients, Bowen could see the struggle to be a separate self. Nature had built in sensitivity to manage short adaptations to environmental demands. That sensitivity, known as anxiety, contained each individual's capacity for more or less factual assessment in responding to threats. Anxiety is an alert to the biological base of an inner emotional guidance system. That system includes automatic responses of fight, flight and freeze and the behavioral markers of agitation (acting out, blaming, demanding rights, withdrawal, physical symptoms, emotional upset, inability to focus, deliberate distraction to avoid important matters, self-medicating with alcohol or drugs and even the calming that can come from caretaking another). The concept of an emotional system replaced the Freudian concept of the unconscious fitting with the human as a part of all living things. This understanding continued to be developed as new information emerged about the family at NIMH.

Courtship, nature's built in response to the need for reproduction and bonding in earliest days with a new baby are examples of dependencies that serve to sustain life. Symptoms are a strength, an individual's effort to adapt and to bind reactivity to free continuance of capabilities. All of us experience a range of functioning that is dependent on what we need and what we do. In a prolonged symbiotic relationship, functioning varies with the difficulty of sustaining positive closeness. Wanting more or less closeness creeps in and cycles can occur.

Basic self can be developed in anyone with directed effort. A threat response emerges when there is a challenge to the functional state to do more, do it differently or do what one does not want to do, giving opportunity to decide for self.

Symptoms were evidence of an individual's own effort to halt the regressed interdependency—a mark of imbalance between individuality and togetherness. If the environment, that is, other people, interfered with the close relationship and symptoms returned, it could take hours or days to get back to the non-symptomatic state (Bowen, 1957a). Bowen discerned that the symptoms themselves were not a result of the schizophrenia or alcoholism; the symptoms were related to the struggle within self toward independence. This struggle within the patient was written up in process notes as Bowen observed it. There we can see the observations that were helping to form the theoretical assumption of a human emotional system.

The capacity to form symptoms is common to all life forms. It is a basic instinct. Symptoms give an organism information on itself and the environment. As an evolutionary adaptation, this builds in options for managing self in the environment that do not need to be permanent. Symptoms are an early alert to pay attention that adaptive capacities are reaching the outer limits of an organism's managing capacities. The ability to develop symptoms, to shift from symptom to symptom and to use symptoms as a way of managing relationships with other organisms further extends survival possibilities. Modern studies on the way the stress response works within an organism and the regulating effects of the environment (including other organisms) on the individual organism illustrate the interconnectedness of individual and environment. The human brain, with its capacity for cortical override of the limbic system, offers another alternative to methods of adaptation, such as symptom development. Neurofeedback equipment can now show the brain's current capacities while bringing change in the brain. Increasing awareness of one's own level of reactivity can give an individual a choice in how to respond. This choice can be a superior response to what the automatic, no-thinking-needed response might otherwise be.

In a letter to an old friend in 1978, Bowen mentioned his lifelong efforts to manage his own autonomic nervous system.

> The researchers are doing things I was doing intuitively in the 1950s. I was doing things to train my autonomic nervous system back in the 1930s and 40s. I still do not wear glasses.
>
> *Bowen, 1978, 1*

Moving beyond maternal deprivation

The greatest reaction from patients with alcoholism and with schizophrenia came not from being deprived of maternal love, but from the staff's "… forced overgiving and compliance" (Bowen, 1960b). Interestingly, in the clinical symbiotic relationship between Bowen and patient, the only intervention that yielded any change was complete passivity by the clinician. For example, the mild indifference in a remark such as "What is the rush to grow up?" proved effective (Bowen, 1957a). A similar observation occurred with the families in the NIMH research project. When the parents were more invested in each other than in the child, the child improved.

At NIMH, when writing of his Menninger experiments, Bowen reflected on how the regression patients themselves recognized that it was over-attachment, not rejection, that needed addressing.

> 'My mother and I are Siamese twins. I want you to separate us so we both can live.' Another said, 'But doctor, you do not understand how it was between my mother and me. We are so close that the only way for either of us to live is for one of us to die, but if either of us dies, then we both will die.' The same patient, in a later stage, said 'It takes a lot of doing to hold your mother's hand and play baseball at the same time'[1]
>
> *Bowen, 1957d, 16*

> 'When I am upset, I am miserable and I can think only of what she has done to ruin my life. In my calmer periods, I know I was born when my mother needed someone for her own loneliness. It could have been my brother or my other sister if they had been born when my mother needed someone. Now, I would be as helpless without her as she would be without me.' If there was one subject that was spontaneously mentioned by patients, it was comments about an umbilical cord attachment to the mother. They would speak of the mother holding onto them while they were clutching the therapist. One side of the story was told in historical material in the psychotherapy. The exact opposite side was being re-enacted in the relationship with the therapist. Occasionally mothers would come for visits and there would be an opportunity to see a real life enactment of the mother patient relationship.
>
> *Bowen, 1957d, 17*

By 1951, Bowen's third year on staff at Menninger, he was considering the benefits of keeping the patient within the family and doing therapy that included them (Kerr & Bowen, 1988, 370). He was also questioning if treatment would prove more salubrious and efficient if it focused on the origin of the problem rather than on a transference of it to another relationship to correct it. Just go to the source, the parent–child relationship. The basic plan, later implemented at NIMH was already forming.

Moving further toward family

Following the idea of giving the family member back to the family, Bowen had parents of in-patients stay nearby. These visits lasted up to six weeks, and both parent and patient attended the daily patient psychotherapy hour. Better clinical results came with family involvement in the patients' treatment (Bowen, 1956). Bowen requested that a section of the hospital or a cottage on the grounds be available to allow the parent to room in with the patient (Guerin, n.d., 12). The Menninger Clinic administrators refused the request. Bowen then implemented an approximation of that idea. The parent could visit daily, giving partial care to the hospitalized offspring. This effort foreshadowed the arrangements Bowen made in the milieu at NIMH. Observing the actual parent and child could add substance and perhaps solidify an alternative theory.

He no longer thought of prescriptive assessments, psychometric tests and diagnoses to understand psychiatric problems.

> I stopped the routine use of psychological tests 10 years ago and since I have been working more with families, I have not even permitted myself to think in terms of diagnoses and 'what is wrong'. This principle has become so much a part of my entire life that it applies to all my relationships.
>
> *Bowen, 1961*

Two of Bowen's ideas seemed to be difficult for his contemporaries to grasp: (1) The family has innate capacities, given certain conditions, to strengthen itself; and (2) a detached observer is more valuable than an emotionally bonded outsider. This second idea is that a neutral other evokes innate growth. Transference remains within the family. Innate growth, when activated between family members brings longer lasting, more secure change. This is a natural systems idea. While this point is accepted among today's students of the theory, it awaits further confirmation.

Fieldwork for concepts: differentiation, emotional system

Symbiosis and a scale as a way to understand relationships can be found in the literature as early as 1943.

> A specific type of relationship ... between ... psychotic persons .... It seems that they become so necessary to each other that any solution is preferable to dissolution to the relationship ....The overburdening of one quickly displaces the other, too. As a rule, one of the persons involved is dominant, and the · other submissive ....The dominant one seems to be the stronger, and the submissive one the weaker individual ... it is not entirely an apt description, for the dominant character in some ways is very dependent upon the weaker for support in his psychotic beliefs ....There is a balance set up between them, as with a scale. The dominant one does not support the other in the sense that a foundation supports a weak superstructure, for a foundation may still stand when its superstructure is removed. The removal of the submissive person, however, will definitely injure the supposed dominant one. Although the two are dependent and interdependent, they cannot be singly self-sufficient. That is to say that they are in dire need of each other.
>
> *Gralnick, 1943, 320*

Bowen discovered that attempts by one person to withdraw from or force togetherness in a symbiotic relationship disturbed the other person. And when the relationship's attachment was disrupted, symptoms occurred.[2] Bowen thought the return to psychotic symptoms was "an effort at resolution that failed" (Bowen, 1960a, 66).

Improvement could occur within the individual in a relationship in which neither party encouraged change nor threatened loss of the relationship. Treatment directed through the helplessness of the base state produced change toward maturity.

Reviewing the differences in individual theory with his findings, Bowen wrote:

> Psychoanalytic theory was formulated from a detailed study of the individual patient. Concepts about the family were derived more from the patient's perceptions than from direct observation of the family. From this theoretical position, the focus was on the patient and the family was outside the immediate field of theoretical and therapeutic interest …. Individual theory was built on a medical model with its concepts of etiology, the diagnosis of pathology in the patient, and treatment of the sickness in the individual also inherent in the model are the subtle implications that the patient is the helpless victim of a disease or malevolent forces outside his control.
>
> *Bowen, 1966, 148*

Whereas psychoanalysis conceived of psychological growth as a product of insight, natural systems conceived of psychological growth as a product of relationship balance. In Bowen's view, the "medical model re: psychopathology–diagnosis– prognosis–psychotherapy" (Bowen, 1957c, 7) was an attempt by Freud, a physician by training, "to bridge the gap … between medicine and psychoanalysis."

The parent as the cause of the problem was the other serious problem that Bowen identified with psychoanalysis. Two fields, psychiatry and social work, integrated to provide an end run around the conceptual dilemma of blaming parents. In the regression method, the emotional issues observed by Bowen and nurse functioning as the parents were also documented by social workers taking the family history of the biological parents. Bowen's other data, his direct observations of parent/ child relationships, also did not support the hypothesis that parents were the causative agents. Rather, there was an entanglement between parent and child. Bowen observed that this tangle sustained the patient's symptoms when a demand was made in an interaction between parent and impaired offspring. This also held true for the psychiatrist; if he gave or included a demand, the patient would predictably return to psychosis.

The regression experiment showed that psychotic symptoms were fluid. In a regressed state, there were no symptoms. In the intense closeness of the clinical symbiotic relationship, the participants were also unmotivated toward growth. Any efforts by the clinician to promote growth produced clinging. Any effort by the clinician to distance from the intense closeness brought increased helplessness. With any trigger conditions, the patient would be more helpless, clinging and psychotic.

In Freudian theory, the "force" shaping one's life was the unconscious memories from childhood and accessing these could alter the course of one's life. Bowen's new idea was that the forces shaping one's life were those of togetherness and differentiation. A person using one's ability to think in relation to important others could alter the course of that life. By 1988, Bowen succinctly described the change in him while still at Menninger.

> It had finally been possible to combine parts of Freud with parts of Darwin, thru the use of the newly created natural systems theory, not previously applied to the human. The human was different when seen thru the new theory, than had been possible with Freudian Theory. One difference had been the number of people who automatically 'clumped' to form one person in certain situations. There was no way to pursue research in an institution devoted to Freud alone .... Several theoretical changes were present in the new theory. The term 'emotional systems' described 'the clump' (instead of unconscious as used by Freud ... and applied to individuals.) 'Differentiation of self' described the degree to which individuals participated in 'the clump'.
>
> *Bowen, 1988*

With regression treatment, the alcoholic and schizophrenic patients responded similarly, their behavior patterns were consistent between the two symptomatologies. While the therapeutic relationship fit with descriptions of the symbiotic relationships in the literature, Bowen added his ideas that a symbiosis was transferable, observable by others, treatable and replicated in early parent–child dynamics. Factually it took the participation of both partners in a relationship to sustain the symbiosis. Excessive love with demands, not deprivation of love, brought the greatest reaction in both groups of patients. One of the last modifications in his regression work was to use nurse and clinician as representatives of the biological parents, and the important modification was to not encourage the transference (which was then tested with out-patients). The idea of a transferable symbiotic relationship showed that Bowen had moved his focus from the individual or a two-person relationship to a three-person relationship—patient, parent and clinician.

These observations show that functioning rather than pathology was already being thought about. Functioning was not a fixed state; rather, it depended on others in the relationship. Psychosis was a functional state managing relationship demands. "Function is what the organism <u>pretends itself to be</u>. Brain research may eventually have answers. Structure makes a big difference between the BRAIN and the MIND ... Function does not distinguish between the two" [capitalization in original] (Bowen, 1989).

Differentiation was about functioning as a separate self within the family by activating the life force of individuality while in an intense, attached relationship. The concept of symbiosis is consistent with the concept of the human as a biological being. Seventy years after Bowen interpreted the interdependency of family relationships as symbiosis, the same concept is now being stressed in life sciences, providing an opportunity for more integrated thinking. "Symbiosis is becoming a core principle of contemporary biology, and it is replacing an essentialist conception of 'individuality' with a conception congruent with the larger systems approach ..." (Gilbert, Sapp, & Tauber, 2012, 87).

These new theories increasingly understand interdependence within and between species as natural laws of living things, which are being conceptualized with the observation and the absorption of broader data.

> The notion of the 'biological individual' is crucial to studies of genetics, immunology, evolution, development, anatomy, and physiology. Each of these biological sub disciplines has a specific conception of individuality, which has historically provided conceptual contexts for integrating newly acquired data. During the past decade, nucleic acid analysis, especially genomic sequencing and high-throughput RNA techniques, has challenged each of these disciplinary definitions by finding significant interactions of animals and plants with symbiotic microorganisms that disrupt the boundaries that heretofore had characterized the biological individual. Animals cannot be considered individuals by anatomical or physiological criteria because a diversity of symbionts are both present and functional in completing metabolic pathways and serving other physiological functions. Similarly, these new studies have shown that animal development is incomplete without symbionts. Symbionts also constitute a second mode of genetic inheritance, providing selectable genetic variation for natural selection. The immune system also develops, in part, in dialogue with symbionts and thereby functions as a mechanism for integrating microbes into the animal-cell community. Recognizing the "holobiont"—the multicellular eukaryote plus its colonies of persistent symbionts—as a critically important unit of anatomy, development, physiology, immunology, and evolution opens up new investigative avenues and conceptually challenges the ways in which the biological sub disciplines have heretofore characterized living entities.
>
> *Gilbert et al., 2012, 86*

Pulling it all together

Bowen could now consider schizophrenia and alcoholism as symptoms of over-attachment rather than a result of rejection or inadequate mothering. Over-attachment, defined as the intense oneness between two people, was characterized as a psychological symbiosis. The fact that it also occurred in the clinical relationship showed that replication, or a transfer of the symbiosis, could occur in other relationships. Amie Post, Executive Director of the Family Crisis Center of Baltimore County, suggests Bowen's shift from symptom as pathology to symptom as a natural part of an emotional process between mother and child is "essential to be able to get to neutral regarding the two life forces" (Post, personal correspondence, September 2020). This deemphasized the rational brain as it holds true across species and takes emotional process back to biology. This shift also places symptom emergence with the concept of an emotional system.

In Bowen's theory, two life forces are postulated: one of individuality or differentiation and the other of togetherness. Defined as neutral life forces, both are necessary for survival. By increasing one's awareness of these forces, a person increases choices for self. These choices allow one to enjoy togetherness without being stuck in the relationship and to represent self without losing the relationship.

A neutral stance in therapist or nurse in the regression study enhanced positive outcomes in the treatment. When the therapist could be present but passive—not encouraging, directing or emotionally withdrawing—the patient responded with improvement. "The program had been started with a conviction that the schizophrenic patient, just as all living things, has somewhere a potential for growth and that this growth can take place if the necessary conditions for growth are provided" (Bowen, 1957a).

The in-patient treatment protocol implicitly identified those "necessary conditions": togetherness with a nondependent, nondemanding, emotionally present other human being, and separateness, as in I am I, and you are you. With both life forces, the human being can activate innate growth for self.

Bowen discussed his 1949–1953 studies this way:

> The initial work was based on previous experience with individual psychotherapy in schizophrenia. It began with a five-year clinical study in which various members of patients' families also had individual psychotherapy. As the emphasis shifted to include family members, the relationship system between family members came into prominence. Attention was focused on the symbiotic attachment between mothers and patients. Of particular interest was the cyclical nature of the symbiotic relationship in which mothers and patients could be so close emotionally that they were 'emotional Siamese twins,' or so distant and hostile they could repel each other. Characteristics of the symbiotic relationship were incorporated into a detailed hypothesis to explain schizophrenia.
>
> Bowen, n.*d.d*

This hypothesis was the starting point of Bowen's research at NIMH in 1954.

Notes

1 This patient is referenced in the discussion of mother–child reciprocity in Chapter 4, "A family concept of schizophrenia," in his 1978 book, *Family Therapy in Clinical Practice*.

2 Robert H. Dysinger, an investigator on Bowen's project, studied this also. His 1956 paper, **"Acquired hemolytic anemia associated with acute schizophrenic psychosis, a clinical note,"** reported on the onset of physical symptoms when a relationship was disrupted. This paper is available in the Bowen archives at NLM.

References*

*Unless otherwise specified, the works of Murray Bowen included in this chapter are from the Murray Bowen Papers. 1951–2004. Located in: Modern Manuscripts Collection, History of Medicine Division, National Library of Medicine, Bethesda, MD. The Accession, Box and Folder information differs for each one and that is included here.

Benedek, T. (1949). The emotional structure of the family. In R. N. Anshen (Ed.), *The family: Its function and its destiny*. New York: Harper and Brothers.

Bowen, M. (n.d.a). [Draft]. (Acc. 2005-055, Box 3, Folder Working papers on theory).

Bowen, M. (n.d.b). [Draft]. (Acc. 005-055, Box 3, Folder Working papers on theory).

Bowen, M. (n.d.c). [Transcript]. (Acc. 2007-012, Box 1, Folder Interview M. Bowen & K Wiseman Interview I).

Bowen, M. (n.d.d). [Draft: "Family psychotherapy with schizophrenia in the hospital and in private practice"]. (Acc. 2007-073, Box 3, Folder Working papers (book chapter on family psychotherapy and schizophrenia)).

Bowen, M. (1951). [Presentation: Direct gratification of the infantile need in schizophrenia, October 1951]. (Acc. 2006-003, Box 5, Folder Papers, schizophrenia notes, etc.).

Bowen, M. (c. 1956). [Drafts]. (Acc. 2006-003, Box 2, Folder Drafts 1957–1958).

Bowen, M. (1956). [Project—Special seminar on schizophrenia]. (Acc. 2006-003, Box 5, Folder Formulation of 3 east family study project July 16, 1956).

Bowen, M. (1957a). [A psychological formulation of schizophrenia]. (Acc. 2006-003, Box 2, Folder Project tiles).

Bowen, M. (1957b). [A psychological hypothesis of schizophrenia]. (Acc. 2006-003, Box 8, Folder A working psychological hypothesis of schizophrenia).

Bowen, M. (1957c). [Draft: A psychological concept of schizophrenia]. (Acc. 2006-003, Box 8, Folder A psychological concept of schizophrenia).

Bowen, M. (1957d). [Draft, A family concept of schizophrenia]. (Acc. 2007-073, Box 3, Folder Isolated working papers re: Schizophrenia & psychosis: A family concept of schizophrenia).

Bowen, M. (1959). [Letter to Helm Stierlin on September 25, 1959]. (Acc. 2007-012, Box 2, Folder Misc. correspondence).

Bowen, M. (1960a). A family concept of schizophrenia. Chapter 4 in *Family therapy in clinical practice*. New York, NY: Jason Aronson, 1978.

Bowen, M. (1960b). [Outline for proposed report about family research project, July 1960]. (Acc. 2006-003, Box 8, Folder Printed papers in odyssey, Part 1 of 2).

Bowen, M. (1961). [Letter on June 12, 1961]. (Acc. 2007-012, Box 2, Folder misc. correspondence).

Bowen, M. (1962). [Handwritten draft]. (Acc. 2014, box 6, Folder Chapter II theoretical background).

Bowen, M. (1965). Family psychotherapy with schizophrenia in the hospital and private practice. Chapter 8 in *Family therapy in clinical practice*. New York, NY: Jason Aronson.

Bowen, M. (1966). The use of family theory in clinical practice. Chapter 9 in *Family therapy in clinical practice*. New York, NY: Jason Aronson, 1978.

Bowen, M. (1976). Theory in the practice of psychotherapy. Chapter 16 in *Family therapy in clinical practice*. New York, NY: Jason Aronson, 1978.

Bowen, M. (1978). [Letter on December 2, 1978]. (Acc. 2007-012, Box 4, Folder Name redacted).

Bowen, M. (1988). [Origins of fam. theory and therapy in mental health]. (Acc. 2006-003, Box 8, Folder Family evaluation).

Bowen, M. (1989). [Letter on January 31, 1989]. (Acc. 2004-043, Box 4, Folder Picone/ Bobick).

Gilbert, S. F., Sapp, J., & Tauber, A. I. (2012). A symbiotic view of life: We have never been individuals. *The Quarterly Review of Biology, 87*(4), 86–87. https://doi.org/ 10.1086/668166.

Gralnick, A. (1943). The Carrington family: A psychiatric and social study illustrating the psychosis of association or folie a deux. *The Psychiatric Quarterly, 17*(2), 294–326.

Grinnell, F. (2009). *Everyday practice of science: Where intuition and passion meet objectivity and logic.* New York: Oxford University Press.

Guerin, P. (n.d.). [Draft of first chapter of book by Philip Guerin], L. Murray Bowen papers, National Library of Medicine, History of Medicine Division. (Acc. 2007-012, Box 3, Folder Guerin), Bethesda, MD.

Hill, L. B. (1955). *Psychotherapeutic intervention in schizophrenia.* Chicago: University of Chicago Press.

Kerr, M., & Bowen, M. (1988). *Family evaluation: An approach based on Bowen theory.* New York: W.W. Norton.

Lidz, R., & Lidz, T. (1952). Therapeutic considerations arising from the intense symbiotic needs of schizophrenic patients. In E. B. Redlich (Ed.), *Psychotherapy with Schizophrenics.* New York: International Universities Press (pp. 168–178).

Mahler, M. (1952). On child psychosis and schizophrenia. *Psychoanalytic Study of the Child, 7,* 286–305.

Mahler, M., Pine, F., & Bergman, A. (1975). *The psychological birth of the human infant, symbiosis and individuation.* New York, NY: Basic Books.

Nichols, M. P. (2008). *Family therapy: Concepts and methods* (8th ed.). Boston: Pearson Publishing.

No author. (c. 1957). [Draft]. (Acc. 2006-003, Box 7, Folder Family research project, notes and paper drafts).

Oberndorf, C. (1953). *The history of psychoanalysis in America 1953.* New York: Grune & Stratton.

Reichard, S., & Tillman, C. (1950). Patterns of parent-child relationships in schizophrenia. *Psychiatry, 13,* 247–257.

8

A SHIFT TOWARD FAMILY 1949–1953

These years advanced the foundation Bowen was building. It was in the back and forth of his varied positions where Bowen gained knowledge of human behavior and of effective milieus in different settings.

Marriage counseling

Bowen became a consultant to the marriage counseling service, a part of the Menninger Clinic services in 1949. The service was one of two in the country (Johnston, 1954b, 119). There, he saw spouses together as "long as it was 'consulting' or 'counseling' and no 'therapy' was involved" (Bowen, 1966a, 2). He held this position until 1954.

This position allowed him to study marital relationships and was another opportunity to research his ideas in more than one setting. At the marriage clinic, he functioned as the connection between psychiatrists and marriage counselors. His work was to show "… working solutions to some of the nuances of difference between psychotherapy and counseling," and he was a teacher and supervisor of the marriage counseling trainees (Bowen, 1966a). Although the clinicians were using a psychoanalytic approach, Bowen was observing long-term therapy with neurotic-level marital problems, which would contribute to the wide swath of evidence that would make up a family theory. Years later, this background contributed to the formation of the American Family Therapy Association (Bowen, 1979).

Exploration with out-patients

The regression experiment was a year into its work when opportunities opened within the community. Bowen was part of a grassroot citizens group in 1949 that formed Shawnee Guidance Center in Topeka to offer local psychiatric services. "I

DOI: 10.4324/9781003027287-9

was the Director of the City-County Neuropsychiatric Clinic in Topeka, which I did one day a week as a donation of the Menninger Foundation to the City-County Welfare Clinics" (Bowen, 1964a, 2). The group of community leaders worked with Bowen to establish community mental health services (2). He became Shawnee's first Clinical Director.

The out-patient clinic, the marriage counseling service and Shawnee Guidance Center were three locations where Bowen directly saw family members for individual psychotherapy, patients for individual psychotherapy and couples for counseling while overseeing residents working with these constellations.

The Shawnee Guidance Center is where Bowen implemented his non-mothering approach to treatment, the opposite of his in-patient treatment approach. By examining his ideas in different theoretical environments, Bowen could challenge the basic tenet that growth was innate given identified conditions and further refine his theoretical assumptions. By going up a level of functioning to nonhospitalized patients, he could verify if his observations would hold up beyond those for severe schizophrenia and alcoholism. If yes, then the method had great applicability.

His out-patient explorations ran parallel to the in-patient program from 1951 to 1953, when he no longer added patients to that program. Bowen tried this method with a few in-patients. "The patient was asked to make decisions about himself, giving him alternatives when possible and parents were asked to make the decisions at crucial periods. This made the parents much more a part of the treatment program" (Bowen, n.d.a).

Bowen's trajectory of orienting toward family came with continuing modifications in treatment as a result of his new understandings of symbiotic relationships between parent and child. The way to help a child was to give mother and child each a therapist. By seeing multiple family members, including spouses, the family relationships came into view. The observations formed the base of a hypothesis suggesting that the beginning of schizophrenia was connected to family relationships (Bowen, 1966c, 27).

> The individual psychotherapy with multiple members of the same family started five years [1949] before the formal research study, [at NIMH] ….This was the first shift toward a family orientation. Clinical evidence of the larger family phenomenon was present at that time but obscured from view by the theoretical thinking, which made it difficult to see beyond the individual. Attention was focused on the symbiotic relationship, which had already been described in the literature as an extension of individual theory.
>
> *Bowen, 1964b*

Bowen's questions about the interconnectedness within generations of a family about emotional illness supported his curiosity about family (Bowen, n.d.b). It was the out-patient effort that proved most useful.

Shawnee Guidance Center, non-mothering approach

In the years 1951–1952, Bowen's scientific inquiry broadened to include people with less severe problems who could get along without hospitalization. He included family members in this work, believing it benefitted the impaired offspring. He recognized that change in a family offered a substantially different understanding of human actions.

If using the doctor and nurse as substitute parents gave insight into actual family relationships, could Bowen check if his own personality played a part, or if the method itself was the means to positive change? A refined treatment method of "non-mothering" was an effort to explore the possibility of keeping the problem centered in the family while relating to a representative of the family, the patient. This revised approach investigated the universality of the theory by using medical residents to serve as research assistants and work with nonhospitalized people who had psychotic level problems. The results of the residents would either support Bowen's understanding and approach or offer new clues for integration or ideas for further modification. It was theory building.

Training the medical residents emphasized how to relate in a way that avoided transference as a therapy method. The residents were unlearning the basics of psychoanalytic treatment. Bowen assembled a team of field workers to report back to him and expanded his research field for data gathering. A case could be made that the training program for medical residents at Georgetown University, which morphed into the Georgetown Family Center's larger student body, is a continuation of this data collection, expanding the potential to find exceptions to natural systems theory. Bowen built in a self-correcting mechanism for his theory building. Students under his tutelage in later years could make a research project of their own lives.

At the guidance center, special emphasis was on developing a "non-mothering attitude" (Bowen, n.d.c, 20). Bowen understood attitude as the complex feelings, beliefs and disposition to respond in a certain way. He knew from the in-patient experience that a mothering attitude "was quite as effective, though not as fast, as the actual mothering in establishing a symbiotic therapy relationship" (20). Bowen instructed the residents to avoid being a substitute parent that offers a corrective emotional experience. They were to engage and show interest in the person, to discover and recognize the individual's ability to manage self in his or her own family. Residents were to be useful to the patient as he or she worked to change personal responses while letting the patient make the choices. The resident could discuss roadblocks or incomplete assessments without giving advice.

Bowen also asked the residents to work on their own selves for greater efficacy in the treatment relationship. It was an opposite approach to regression. They were to avoid any functioning in a caretaking or mothering way. This meant paying attention to their own reactiveness. Bowen trained residents how to avoid any behavior known to increase the symbiotic attachment.

Here, residents recognized the adult capacities of the patient. This meant hearing beyond complaints, delusions, hallucinations and faulty assumptions. When the person related something from the mature part of self, it prompted support and discussion. Failure occurred when the residents did their best to portray this non-mothering attitude but underneath had "overprotective thinking" (Bowen, c. 1957, 7). The patients sensed their pretending. It took time for the residents to integrate the intellectual understanding with the emotional resonance of the patient's dilemma.

> Have a non-mothering attitude – removing any mothering action, or mothering thinking about the person, any infantilizing thinking toward the patient-'Poor unfortunate impaired, helpless, pitiful, deprived person with all those unmet needs for nurturing and understanding' (any thinking along these lines sets up the possibility of a dependent relationship).
>
> *Bowen, 1957*

The outlook of the clinician was crucial. The training rationale emphasized that developing this attitude brought better results. In reflections preparing for his chapter in the 1988 book *Family Evaluation*, Bowen elaborated.

> Do the things that establish the person and the therapist as separate, autonomous entities. Find a way through the opposing bids from the patient to be the patient's parent and to treat him or her as an adult. Begin pointing out relationship distortions and avoid having them acted out. The therapist is not the parent or a parent substitute (ex: if a patient asks for money, don't interpret, rather 'I'm your therapist not your father'). Work toward establishing autonomy of self by avoiding projection or interpretation. One can only know what [the] other thinks and feels if the individual expresses it, avoid such comments as 'I understand, I think that you.' If the person improves, avoid 'I am pleased or proud.' Reflect the person by 'You sound pleased.' Avoid comments that make the therapist the expert 'You feel psychotic' as it was known that the person would accept the comment as if it were a fact. Give choices that require the person to decide which way he or she will go: rather than 'Here's an explanation for how you are' offer 'Here's what I think, others hold opposite views' or 'some people get better some don't.' Relate only to the adult in the patient. Make explicit the psychotherapy contract-patient states what he wants from therapist, therapist is then explicit with what he will do and what he will not do.
>
> *Bowen, 1987a*

To help the development of self, residents left the patient's unmet needs and decisions to the patient, just as in the in-patient work. The client was to decide for self the extent of attention wanted (half-hour or hour-long appointments) and frequency of appointments (one or more times a week). The patients chose less time and fewer appointments than what the resident would have suggested; progress was more than residents predicted (Bowen, c. 1957, 6).

Bowen knew attitudes transferred between people. How one conveyed one's self to another and how one responded to another shaped the relationship. This knowledge was the foundation of a more neutral approach to bring change toward increased self-responsibility in both clinician and client. Interestingly, it produced increased contact between parents and child.

Family involvement was fundamental to this program, the thinking being to give the patient back to the family. Recognizing/expecting that the family had strengths undermined the hierarchy of superiority often inherent in treatment. The clinician was a source of information and assessment for the family's own ideas, not the expert who had the answers. By using the clinician for support to draw upon only when needed, families could select the best options within themselves to advance their own family. Parents were more active, and they could make decisions for their adult child ordinarily made by the psychiatrist.

> The orientation had evolved to the point that it was very similar to the treatment plan in general use by child psychiatrists for many years.
>
> *Bowen, n.d.a*

Hints at emotional processes among three people arose in these observations. The term triangle first shows up in work from January 1953 (Bowen, 1953). Later writings used the term triad.

Information on the participants in the program comes from Bowen's writings about the program. Residents saw five out-patients with good results using the non-mothering method. The participants were seriously impaired people with enough inherent strength to circumvent in-patient hospitalization. As with the anaclitic regression efforts, the sample size was small. But the positive results convinced Bowen to consider this a treatment possibility with more controlled conditions, which he did when designing his project at National Institute of Mental Health (NIMH). His intention was to get viable results from this study. Bowen noted to his "surprise" that disturbed people stepped up to accepting responsibility for self, and at the capacity they showed to "maintain an 'observing ego' distance from the psychotic symptoms and to do meaningful psychotherapy" (Bowen, 1962).

Moving forward

The out-patient experience brought an end to the hospital regression effort.

> Experience with the regression focused the treatment program on the help-lessness in the patient rather than on the symptoms. There was indication too that the motivation for growth was in the patient (as seen when the patient was symptom free) and that well intended efforts to be helpful could block the patient's own initiative and motivation .... The regression program, as a

treatment method, was abandoned. Principles learned in the program were adapted to out-patient psychotherapy.

Bowen, n.d.a

The transition from the mothering approach to an opposite approach had a big impact on theory development. At its base was an understanding of two principles. The first principle predicted the discovery of the family as an integrated unit. It put the problem back with the family even if only one family member displayed the symptoms (a systems idea). Involving family in treatment acknowledged this principle. The second principle, following from the first, was that change from within the family was longer lasting when the clinician was a resource, not a solution. No longer was clinical transference the primary change agent.

Bowen described the divergence between each approach in later writings and summarized the advance of his theory's development.

> The term 'mothering' at the beginning of the regression program, had been used to refer to the kinds of personal attention a mother might provide for a child. Later, the term 'mothering' referred also to the mothering attitude, which included the kinds of protective feelings, attitudes and thoughts a mother might have about a child. The term 'non-mothering' referred to the absence of mothering attention and attitudes that would be consistent in the relationship of a mother to a child. Non-mothering referred to the kinds of attention and attitudes that would be consistent in the relationship between a mature mother and an adult child, or consistent in the relationship of one adult to another adult. The term 'non-mothering' was chosen because it was the opposite of the mothering attention which was stressed throughout the regression treatment.
>
> *Bowen, 1957*

Bowen's use of the non-mothering method in a freestanding clinic beyond the Menninger complex contributed importantly. Here, families who lived locally could be involved without concern for lodging or time off from work.

Bowen's experiment with the non-mothering approach shows theory building in vivo. It was a decisive moment bringing qualitative change in his development of theoretical concepts. While the regression effort explored the potential negative influence of the mother it instead yielded basic components of a theory, the process of separation and a human inner guidance system. The non-mothering approach explored the potential influence of the intense attachment between the parent and child and a treatment program to resolve that to more equally balance the relationship between parent and child. Bowen writes that his out-patient effort "completed the change that went from one theoretical concept to another and then to a different treatment plan" (Bowen, 1957).

Bowen defined how each program anticipated the NIMH research effort. The non-mothering approach was the final step in this eight-year evolutionary process. The final stage was to discourage intense relationships with any family member, to leave the intensity of the process within the family, and to relate peripherally and supportively to various members of the family unit. The more the family was involved in all levels of decisions and actions in the patients' treatment, the better the clinical results.

Bowen, 1956, 2

How Bowen used his personal principles to advance his interests

All the therapy programs he used involved the giving of love. Bowen carefully defined this term as "time and attention" (Bowen, 1995, 24). A 1944 article on "Psychotherapy and 'Giving Love'" by Nathan Ackerman describes the concept of giving love. It was not a technique. It required skillful assessment in the clinician to know when a patient can receive love. And it never replaces the love lost in childhood within the parental relationship (Ackerman, 1944, 12). Ackerman is an important figure in the history of psychiatry and the family. He was a resident at Menninger in the early 1930s, when there were only one or two residents a year and he was another of the earliest psychoanalytic thinkers about family. He joined the staff at Menninger and in 1937 published "The Family as a Social and Emotional Unit" (Ackerman, 1937).

In that paper, he developed important ideas of the family as a biological, psychological and social unit in which early childhood experiences in that unit could influence relationships long into the future. He referred to the family as a unit and described a family emotional system: "The emotional tone which governs the relationship between any two persons of a family has a development peculiarly its own, but is continuously influenced in its course by the emotional relationships of all other persons in the family" (Ackerman, 1982, 156). Some writers consider Ackerman to be the revolutionary who first brought families under direct observation (Zawada, 2015, 27).

Bowen heard of Ackerman in 1946 at Menninger. By then, Ackerman, no longer at Menninger, was seeing couples and their impaired child in psychotherapy using psychoanalytic principles directed to the family and on the fringe of psychoanalytic practices (Brewster, 2015, 11). Ackerman heard of Bowen's research at NIMH, around 1955, and began writing to him (Bowen, 1988a, 1).

Interestingly, in his 1944 article, Ackerman warned against the interventions Bowen made in his regression approaches, warning of the same complications in therapist and social worker relations that Bowen found. Bowen and the nurse were both active with the patient in the same setting and Bowen used the social worker to compare the hospital experience with the family's own reports. The social worker was an important data source and connection to the family. Ackerman warned of "unconscious identification" (Ackerman, 1944, 17) and "unconscious competition with the child's mother" (17). Bowen designed his program at Menninger so that the locus of control remained within the patient, not the therapist and nurse. Ackerman described problems with excessive giving (18) and Bowen also observed this. But Bowen amended the concept by giving love with no demand. When this treatment was applied, a deeper awareness emerged of the two life forces, togetherness and separateness, both necessary in different contexts. This extended the understanding that "forced mothering," a variation of excessive giving, brought a return of symptoms and suppressed natural motivations for growth.

This led Bowen to create a treatment contract, to keep the exchange between patient and treating staff based in reality. Staff were present but making no demands. The difference between Ackerman and Bowen's observations was that Ackerman was using Freud's theory to make assumptions, while Bowen was using direct observations to model his theory. Bowen found that excessive giving in human mothers complicated matters. Primatologist Jane Goodall saw this in nature when she described excessive giving in the mother chimp named Flo who was over indulgent with her last offspring (The 'F' Family, 2017).

In Bowen's varied explorations, each experience contributed to his theory building. His careful attention to words and their influence on the therapeutic relationship is another illustration. In a letter written in 1966 to a former patient at Menninger, Bowen told a story from 20 years earlier. He used ordinary descriptive words instead of psychiatric labels of "sickness." Then he watched to see if the language made any difference. It did. He was learning the potential of activating actual strengths in people. The power of "sick words" and the impact of the environment both contributed to Bowen's understanding during his time at Menninger (Bowen, 1966b). "The principles learned in Topeka, about resolving intense relationships with impaired people, eventually became the baseline for my concept about 'differentiation of self'" (Bowen, 1985, 3).

A larger context

Curiosity both in the family and living systems was increasing within the intersection of public policy and family research. William Menninger was one of the leading founders of the Group for Advancement in Psychiatry (GAP), in May 1946. The group was an exercise for more professionalism, efficacy for treatment, advancements for research and modernization of training within American psychiatry (GAP History, 2019). Menninger intended it to provide a supportive platform for the trainees in the new school of psychiatry (Schlesinger, 2000). Bowen was a member of this organization and the Menninger initiatives aligned or perhaps even shaped his wish in later years that his contribution to psychiatry stay within medicine. It was five to six years before the Committee found field psychiatrists who were interested in the family (Bowen, 1975, 287).

NIMH became an official entity in 1949 (NIH, 2017). In 1951, the National Council on Family Relations published *Family Relations: Interdisciplinary Journal of Applied Family Science* (2020). As one reader said, "The world around Bowen crackled with interest in the family" (A. Post, personal communication, August 23, 2020). Murray Bowen is not the only researcher who placed the human family within biological systems. His daughter, Joanne Bowen, a research professor in the Department of Anthropology at The College of William and Mary writes that Julian Steward, an anthropologist interested in an evolutionary model of adaptation by societies to the physical environment, working during the same years Bowen was developing his theory on the family, understood the deep connection between humans and the earth.

> Bowen was clear that a disconnect from the land would set the emotional stage [for continued human existence], and Steward was clear social relationships- the system through which families connect with the earth – count. Together they demonstrate [that] a connection with the land and social relationships are essential for sustaining life on this earth.
>
> *Bowen, 2020, 310*

Biologically based systems thinking in anthropology organized itself around a family's tie to the land.

Bowen found that discussing a new theory of human adaptation in 1952 prompted never-ending debate. The possibility occurred to Bowen that it might be necessary to leave Menninger to pursue these studies. His final control case to receive credit as a psychoanalyst required a commitment to stay at Menninger until he finished the case, possibly several more years. It is telling that Bowen put this control case on hold in 1952 (Bowen, 1966b, 3).

A crossroads

Now a personal life choice arose. Though it meant he was on his own, as his ideas were difficult for others to accept, Bowen pursued them without concerning him- self about their social acceptance. In extending his ideas to out-patients he was now well into a "solitary effort" that went toward "a more scientific theory of human adaptation" (Bowen, 1988b, 2).

It was during his out-patient efforts that Bowen concluded extending Freudian theory was not a useful undertaking for shoring up the grand possibility of a factual theory based on science. This allowed for what he termed his first actual progress. The in-patient regression work and the out-patient efforts, begun in 1949, led to nodal changes in his thinking by 1953.

> There appeared to be no way to modify the theory (Freudian) with a few simple contributions ....Finally came a new idea. One does not create a new theory from errors about the old. It required some kind of systems theory to handle all the variables. Existing systems ideas had already been defined, but thinking applied to inanimate life. A natural systems idea was created to fit

precisely with each element of evolution. The new theory was formed-Freud, + Darwin + natural systems to combine the two.

Bowen, n.d.b

This supports the argument that Bowen's embryonic theory began at Menninger in the late 1940s and early 1950s. He hypothesized that severe illnesses such as schizophrenia and alcoholism were products of a disturbed relationship process. They were treatable in a milieu with real family members and a particular method of treatment. Bowen defined hypothesis this way: "The term *hypothesis* [italics in original] describes an educated guess about a tiny piece of the total puzzle" (Kerr & Bowen, 1988, 352). To do the research he wanted, Bowen began a search for places to explore his hypothesis in clinical operation. His deep confidence in the validity of further exploration shows in this later life reflection, "By 1951 I believed a viable science of life had to somehow be in harmony with the sun, planets, tides, seasons, reproduction, and with the earth itself" (Bowen, 1986, 5).

In that same year, 1951, Bowen presented a paper on "The Effect of Mobilization on Children and Youth" at a University of Wisconsin conference. Several times, the presentation referenced cycles in national emotional functioning in the United States over the past 20 years (Bowen, 1951, 4–9). The point he emphasized was that the seriousness of the trauma did not leave the most indelible mark, rather how the family managed the trauma made the important difference. Children followed their parents' lead in adapting to crisis. He cited the work of Anna Freud with English children during World War II. He referred to symptoms as "restitutional processes." Symptoms served as an alert and a preservation mechanism. As an example, he described the observations in regression of a person acting in response to a perceived demand within self or from others. After discussing developmental periods in a child's life, Bowen gave two case studies. In these, he emphasized the importance of openness between husband and wife.

In the case scenarios he presented, the paper is a curious mix of the influence of Freudian thinking and ideas that foreshadowed the concepts he would develop at NIMH. These included a deeper understanding of the family unit (a term he used often in the paper), marital conflict, the transmission of parental anxiety to a child and a societal emotional process.[1]

> The greatest gift that mankind has to bestow on a member of the human race is the gift of reasonably mature parents … the average maturity of our society is an extremely variable thing and it seems to run in cycles.
>
> *Bowen, 1951, 29*

Alongside Bowen, American psychiatry was developing more structured standards and a movement toward naturalistic observations. The first edition of the *American Psychiatric Association Diagnostic and Statistical Manual of Mental Disorders* (DSM) was published this year, in 1952 (APA, 2020). Bowen, at Menninger, tried to communicate to others that psychiatry might be better informed by evolutionary theory than by using Greek plays or interpretations of people's stories and symptoms that focused on the individual. He found that "Topeka was 'mecca' for the traditional, and open questions about mecca were not popular there. My head stayed on course in spite of opposition. Eventually it produced a new way of thinking about human problems" (Bowen, 1987b).

Moving toward a more scientific theory

Psychoanalytic theory in the early 1950s considered psychological problems to be a product of relationships, and the ability to form a transference meant psychological problems were resolvable in a relationship. The first relationship in anyone's life is the mother–child relationship. Most often, the mother has the longest and most intimate relationship with a child, beginning at conception and continuing through development. The emphasis on a family relationship did not directly translate into the practice of therapy. Instead, it was how the patient related to Bowen reflecting that relationship that was the source of the idea for resolving psychological problems in those original relationships. What Bowen took from this was that "psychoanalysis ... has concepts about the ways one life influences another" that, in a manner of thinking, started the family movement (Bowen, 1971, 184).

Early in his psychoanalytic formulations, Freud considered the parents in the origin and emergence of emotional illness. However, though some work with families occurred, it was largely absent for nearly 50 years. It was in the 1950s that theoretical conceptualizations of family merged with treatment methods. It was in 1957 that these various researchers publicly presented on their work (Bowen, Dysinger, Brodey, & Basamania, 1957).

Individual therapy dealt indirectly with the relationships between generations of a family while addressing a hypothesized pathology within the individual. If one blamed the parent for the problem in the child, then it made sense that a therapeutic relationship with a healthy therapist outside the family could heal this damage. But considering the human family as a biologically determined group that best ensures offspring survival involves a different level of conceptualizing. If one believed that nature determined the family to be the best location of health for a child, then strengthening the parent's capabilities, without blame, makes sense for the generations to follow.

The assumption that the most important person in a child's life was a psychopathological parent, unconsciously hurtful to the child while consciously trying to be helpful, presented a conceptual problem for a theory of the human that is based on evolution. In nature, a parent normally assists the offspring in growing. Blaming parents left the offspring adrift, without resources or the strengthening and stabilizing possibilities that can come from the most significant people in a person's life. The assumption that someone outside the family needed to take over parenting and guidance toward maturity presented a therapeutic loophole. Bowen wanted to put the person back in the family. It was possible that serious emotional symptoms in an individual benefitted from the innate strengths of the family (Bowen, n.d.d). This agreed with nature and directly brought family relationships into therapy.

Menninger's was interested in proving Freudian theory. Faced with competing paradigms, Freudian theory and an approach to the human as an evolutionary being, Bowen continued on his course.

From his readings, clinical practice and observations of others' clinical practice, he created a theoretical baseline of human behavior. He incorporated facts that had a solid foundation in Freudian theory but, crucially, added ideas with a foundation in evolution.

A more factual theory, with differentiation of self (replacing id, ego and superego) and the emotional system (replacing the unconscious), came from his research at Menninger. Bowen had an ability to track a person's feelings. Perhaps all analysts do. Bowen understood feelings to be one part of a tripartite emotional system. The other parts are the deep emotional responses at the level of instinct built in for survival, and the human brain assesses, regulates and offers choices. Feelings are the bridge between those two parts. In 1987, Bowen wrote, "In my own forty years of theory building, I have carefully avoided the notion of the unconscious and replaced it with automatic instinctual action" (Bowen, 1987c).

This understanding of an emotional system as a substitute for an unconscious served the purpose of Bowen recognizing the ability within each person to override that emotional system, at least to a small degree, through activation of their own intellect. He called this differentiation, a process that begins at birth.

Bowen observed and contextualized what he saw differently from many of his contemporaries, all the while living and functioning in the traditional world of psychiatry. His incremental but profound shift in his theoretical base was one part of his life. It is helpful to remember that concurrent with his forming new concepts, Bowen was actively getting the credentials he needed to complete his psychoanalytic training. From 1949 through 1953, he had three psychoanalytic control cases and seriously considered more personal analysis for himself (Bowen, 1959). He had active involvement with psychoanalysis, was under the supervision of Rudolph Ekstein at the Topeka Institute of Psychoanalysis, 1948–1954, and comported himself professionally within the Menninger Clinic (Bowen, 1955, 1). He held to the ideas found in Westley, Zimmerman and Patton's work on how social innovation occurred. "Intractable problems can be solved" (Westley, Zimmerman, & Patton, 2006, 6), and Bowen knew human behavior was complex in a way Freudian theory did not acknowledge.

Bowen left a record of finding his way to new understanding while practicing Freudian theory. The method Bowen defined for his multidiscipline literature review served as a template for his ongoing research investigation. Beyond requiring consistency with scientific studies of evolving life and examples from the animal world, Bowen formed theoretical assumptions to further his studies. His path to stating his theory publicly took 20 years. By 1966, he described his method of moving toward science in forming a theory and the relationship challenges that accompanied it.

Over the years, I have come out with about four major interlocking concepts that go to make up the family theory. First, there would be the work on one concept and then a shift to bring another into alignment. I have found no way to hurry it. Each fragment of an idea has to be checked and rechecked in hundreds of clinical situations …. For me, the process of knowing what I believe has been a long one. The relationship aspect of this, which presses for conformity, came into focus several years before the family research …. There was relief when I dared to say things different from group beliefs and the only consequence was the predictable flak, the diagnosing of my 'pathology', threats, and temporary rejection by the group. The most difficult aspect of knowing what I believe has come into focus in the attempt to formulate a family theory of emotional illness. The family provides a wealth of theoretical models not possible with individual theory. Application of these models extends far beyond the boundaries of psychiatry. My thinking system never lets go of the problem. Some issues, unclear in the waking hours, have been clarified in dreams. Progress has been in stages. During some periods, I would be crystal clear. Then would come new data that did not fit and a period when it all seemed vague and muddled. In later years, I have not been bothered by attacks from the relationship system except when it consumes time needed for the primary goal. Some attacks can be used to clarify one's own position.

Bowen, 1966d

The observations in both the in and out-patient explorations at Menninger changed Bowen and how he practiced. He made personal efforts to act less on compassion and other common human responses, such as empathy and sympathy for the plight and suffering of others, and he tried to surmount pity and even anger. This created a growth-producing clinical environment that left relationship dependency in the family. He considered being present and engaged to be a more effective alternative to acting with such individual reactions. This was part of the gradual change away from transference (Bowen, 1973).

The environment also shaped Bowen. Better clinical results came with family involvement in decisions and the patients' treatment, resulting in basic changes to how Bowen practiced. He no longer thought about tests and diagnoses to understand psychiatric problems.

In 1951, when Bowen applied to the research department for a designated cottage on the grounds for relatives to live in or stay for extended visits, Menninger rejected the proposal. By 1952, he included fathers in out-patient sessions while keeping the focus around the mother/child symbiosis (Guerin, n.d.). In these years, exploration of families occurred through in-patient, out-patient and marriage counseling.

In his seventh year at Menninger's, he had already navigated the day-to-day world of a residency; practiced as a psychiatrist with in-patient cases; saw out-patients along with families; experimented with and found replacement understandings for id, ego and superego and the unconscious; saw couples at the marriage counseling center which expanded his awareness of that subset and supervised residents who were seeing out-patients at Shawnee Guidance Center using a new treatment method and whom he taught to activate inner strengths and to reduce dependence on transference. In each of these positions, Bowen refined and single-mindedly advanced his knowledge on a substantive path to a more scientific approach. He traveled his own path, seeking facts to create a more solid theory. Families were increasingly important to his studies.

Note

1 This paper is available on The Murray Bowen Archives Project website at http://murraybowenarchives.org/published-papers.html.

References*

*Unless otherwise specified, the works of Murray Bowen included in this chapter are from the Murray Bowen Papers. 1951–2004. Located in: Modern Manuscripts Collection, History of Medicine Division, National Library of Medicine, Bethesda, MD. The Accession, Box and Folder information differs for each one and that is included here.

Ackerman, N. (1944). Psychotherapy and 'giving love', psychiatry: *Journal of the Biology and Pathology of Interpersonal Relations*, 7(2), 129–137.

Ackerman, N. W. (1982). The Family as a Social and Emotional Unit. In D. Bloch & R. Simon (Eds.), *The strength of family therapy: Selected papers of Nathan W. Ackerman* (153–158). New York: Brunner/Mazel ("The Family as a Social and Emotional Unit", originally published in *Bulletin of the Kansas Mental Hygiene Society*, 12:2 1937).

American Psychiatric Association. (2020). *DSM history*. Retrieved July 1, 2020, from American Psychiatric Association, Psychiatrists: www.psychiatry.org/psychiatrists/practice/dsm/history-of-the-dsm.

Bowen, J. (2020). Anthropological contribution to the study of the human family. In M. N. Keller & R. J. Noone (Eds.), *Handbook of Bowen family systems theory and research methods: A systems model for family research* (pp. 301–320). New York: Routledge.

Bowen, M. (n.d.a). [Schizophrenia and the Family]. (Acc. 2007-073, Box 3, Folder Working papers from NIMH project re: schizophrenia).

Bowen, M. (n.d.b). [Draft of Honolulu paper]. (Acc. 2006-003, Box 7, Folder Family relationships in schizophrenia).

Bowen, M. (n.d.c). [Draft]. (Folder Working papers from NIMH project re: schizophrenia).

Bowen, M. (n.d.d). [Origin of the family movement]. (Acc. 2007-073, Box 3, Folder Bowen working papers).

Bowen, M. (1951). [Paper given at "Children and Youth in a World in Crisis", June 27-29, 1951, University of Wisconsin]. Copy in possession of Bowen family, Williamsburg, VA.

Bowen, M. (1953). [Transcription of 227th hour of a psychoanalysis, January 5, 1953]. (Acc. 2006-003, Box 3, Folder Dr. Bowen file no. 14,226).

Bowen, M. (1955). [Letter to Director of Field Services, University of Wisconsin on June 24, 1955]. (Acc. 2006-003, Box 4, Folder Correspondence).

Bowen, M. (1956). [Draft of the family and schizophrenia, January 1956]. (Acc. 2007-073, Box 3, Folder Bowen-NIMH-working papers incl. proposal for project).

Bowen, M. (c. 1957). [Therapeutic and theoretical gains from experimental regression programs]. (Acc. 2007-073, Box 3, Folder Working papers from NIMH project re: schizophrenia).

Bowen, M. (1957). [Draft. A psychological concept of schizophrenia]. (Acc. 2006-003, Box 8, Folder A psychological concept of schizophrenia).

Bowen, M. (1959). [Letter to Rudolph Ekstein, PhD on October 12, 1959]. (Acc. 2004-013, Box 4, Folder PSA Institute).

Bowen, M. (1962). [Part I theoretical background]. (Acc. 2014-034, Box 6, Folder Chapter II theoretical background).

Bowen, M. (1964a). [Letter on June 27, 1964]. Copy in possession of the Bowen family, Williamsburg, VA.

Bowen, M. (1964b). [Intrafamily dynamics in emotional illness]. (Acc. 2004-013, Box 2, Folder Ggtwn conf. Jan 1964).

Bowen, M. (1966a). [Therapist's pathways to family psychotherapy]. (Acc. 2009-013, Box 6, Folder GAP Mar, family therapy questionnaire).

Bowen, M. (1966b). [Letter to Menninger patient on November 28, 1966]. (Acc. 2007-073, Box 1, Folder patients—hold 20 years).

Bowen, M. (1966c). [Explanation of identifying data, section I, 1966 in family therapy questionnaire]. (Acc. 2009-013, Box 6, Folder GAP).

Bowen, M. (1966d). [Partial letter to Don Jackson]. (Acc. 2005-055, Box 1, Jackson, Folder Jackson, Don).

Bowen, M. (1971). Family therapy and family group therapy. Chapter 10 in *Family therapy in clinical practice*. New York, NY: Jason Aronson, 1978.

Bowen, M. (1973). [Letter. November 23, 1973]. (Acc. 2007-012, Box 2, Folder misc. correspondence).

Bowen, M. (1975). Societal regression as viewed through family systems theory. Chapter 13 in *Family therapy in clinical practice*. New York, NY: Jason Aronson, 1978.

Bowen, M. (1979). [Letter April 7, 1979 about AFTA]. (Acc. 2004-043, Box 5, Folder Letters from M. Bowen re: ideas).

Bowen, M. (1985). [Letter to the social worker on Bowen's team at Menninger on July 4, 1985]. (Acc. 2006-003, Box 2, Folder Correspondence materials).

Bowen, M. (1986). [Presentation at Center for Family Consultation on June 7, 1986]. (Acc. 2003-044, Box 6, Folder CFC – Chicago, IL – June 7, 1986).

Bowen, M. (1987a). [Draft]. (Acc. 2006-003, Box 8, Folder Printed papers in odyssey).

Bowen, M. (1987b). [Letter to a former client on July 31, 1987]. (Acc. 2007-012, Box 2, Folder Death & funerals).

Bowen, M. (1987c). [Letter to Edward Bruce Taub-Bynum, PhD on June 1, 1987]. (Acc. 2007-012, Box 1, Folder Interesting letters).

Bowen, M. (1988a). [Letter on August 29, 1988]. (Acc. 2004-013, Box 4, Folder AFTA-special 10th anniversary program).

Bowen, M. (1988b). [Origins of family theory and therapy in mental health]. (Acc. 2006-003, Box 8, Folder Family evaluation).

Bowen, M. (1995). A psychological formulation of schizophrenia. *Family Systems Journal*, 2(2), 17–47.

Bowen, M., Dysinger, R. H., Brodey, W. M., & Basamania, B. (1957). [Paper "Study and treatment of five hospitalized family groups each with a psychotic member" presented at the annual meeting of the American Orthopsychiatric Association in Chicago, Illinois on March 8, 1957]. (Acc. 2007-073, Box 4, Folder Working papers).

Brewster, M. K. (2015). Ackerman, Nathan. In E. S. Neukrug (Ed.), *The SAGE encyclopedia of theory in counseling and psychotherapy*. Thousand Oaks, CA: Sage Publications.

Farreras, I. G., Hanaway, C., & Harden, V. A. (Eds.). (2004). *Mind, brain, body, and behavior: Foundations of neuroscience and behavioral research at the National Institutes of Health*. Washington, DC: IOS Press.

Group for the Advancement of Psychiatry. (2019). *GAP history*. www.ourgap.org/history.

Guerin, P. (n.d.). [Draft of first chapter of book by Philip Guerin]. (Acc. 2007-012, Box 3, Folder Guerin). L. Murray Bowen papers, National Library of Medicine, History of Medicine Division, Bethesda, MD.

Johnston, Q. (1954a). Marriage *counseling services in* Kansas (Vol. 3). Retrieved June 6, 2021, from University of Kansas Law Review: https://digitalcommons.law.yale.edu/cgi/view content.cgi?article=2966&context=fss_papers.

Johnston, Q. (1954b). Marriage counseling services in Kansas and Kansas City, Missouri. *Kansas Law Review*, *3*, 116–129.

Kerr, M. E., & Bowen, M. (1988). *Family evaluation: An approach based on Bowen theory*. New York: W. W. Norton.

NIH. (2017, February 17). NIH Almanac, National Institute of Mental Health (NIMH), *mission*. Retrieved July 1, 2020, from National Institutes of Health (NIH): www.nih.gov/about-nih/what-we-do/nih-almanac/national-institute-mental-health-nimh#:~:text=Kellogg%20of%20Indiana%20University.,the%20first%204%20NIH.

Schlesinger, H. (2000). Musings about my intellectual development. In P. Fonagy, R. Michels, & J. Sandler (Eds.), *Changing ideas in a changing world: The revolution in psychoanalysis: Essays in honour of Arnold Cooper* (pp. 42–55). London: Karnac Books.

The Jane Goodall Institute Australia. (2017, March 22). *The 'F' Family*. www.janegoodall.org.au/2017/03/the-f-family.

Westley, F., Zimmerman, B., & Patton, M. Q. (2006). *Getting to maybe*. Toronto: Random House of Canada.

Zawada, S. (2015). An outline of the history and current status of family therapy. In S. C. Box (Ed.), *Psychotherapy with families: An analytic approach* (pp. 25–34). London: Routledge.

9

TIMING AND CHOICES 1953 TO JULY 1954

Being in the right place at the right time with the right contacts was the thread of fortuity in Bowen's development as a psychiatrist and theoretician. That thread ran through his early family life, medical school, service in the Army and his years at Menninger. It again held true in 1953. In January that year, Bowen wrote to his family that he and Mrs. Bowen were talking about leaving Menninger to build on his new ideas in a research setting. Despite being well established at Menninger, Bowen faced significant roadblocks like having to work surreptitiously with the families of patients. In his letter, he said he had given the Menninger Clinic notice (Bowen, 1953a).

Intervening occurrences in Bowen's life

In June 1953, Karl Menninger presented to the Tuesday Conference at Menninger. While I have not found his presentation, I found a July 1953 memo from Bowen responding to Menninger's talk on expanding thinking about symptoms and illness. In the memo, Bowen described the presentation as "bold" with "far-reaching implications" (Bowen, 1953b, 1). Bowen referenced several points Menninger made about moving toward new understandings and about causality "and the fallaciousness of this notion." Menninger apparently suggested having a mind open to new understandings beyond the then practiced "formula—recognize—foretell—and cure." Bowen's appreciation of Menninger's breadth of thinking was notable.

> It is hard to put together thoughts about your conference …. I was much impressed by your ideas … your thinking … is the best you have ever done … it goes hand in glove with your homeostasis concept which we have all found to be so helpful. I think your homeostasis ideas have done more than anything else to change my own basic orientation to psychiatry and the

DOI: 10.4324/9781003027287-10

treatment of patients since I first heard you begin to speak of it some two or three years ago ... the ideas expressed in this discussion are a logical extension of the homeostasis ideas ... the broad philosophic basis for your ideas would put the concept of illness to a perspective as broad as some of the concepts we are hearing these days from geologists.

Bowen, 1953b, 1

Menninger's presentation sparked Bowen's curiosity about connecting his own ideas with those expressed by Menninger on forces of nature that are neither "bad" nor "good" but connected to processes of "evolution and involution." Bowen wrote of a recent family trip to the Grand Canyon, where he heard a presentation by park rangers.

I was amazed that no one openly questioned the apparent conflicts between these geologic concepts and concrete religious concepts ... the rangers say that rarely comes up. At the Canyon they have the scientific evidence to back up the broad outlines of their ideas. Man probably has to have God to help him bridge the vastness of the concept that would make him little more important than the fossil of some form of life that existed three hundred million years ago .... This might be the kind of thing you have in mind when you proposed the idea that there is no such thing as disease. If it is not disease, then what is it ... it is not so much the label one uses as the way in which we use the label, and this is the point you were making ... I do not think your presentation will be too popular. When you threatened concepts that have been held so long and that have served such a good service in controlling anxiety, then ... the most anxious of men will probably have to pick at the details in an attempt to disprove the ideas.

Bowen, 1953b, 2

He predicted other professionals would have a hard time accepting Menninger's views, which were as yet unaccepted by society or psychiatry.

A pause to rethink leaving

Walter Kass and Bowen had started at the same time in the Menninger School of Psychiatry. Kass was now the psychologist in Bowen's section and the senior supervisor of the post-doctoral program (Bowen, 1957). Bowen considered Kass to be one of the most gifted practitioners at Menninger (Bowen, 1956). He was a sensitive friend and privy to Bowen's views on theory building and his desire to extend that work. Bowen's mothering and non-mothering experiments, which informed his search for clues to understand human life, had laid the groundwork for a nascent, more factual theory that needed formalized clinical application to proceed. It had been a conversation with Kass in 1951 that had prompted Bowen's unsuccessful research proposal for a dedicated cottage on the Menninger grounds to have families "live in or come for long visits" (Bowen, 1966, 2). He credited Kass with helping sort out Bowen's plans to stay or move when Kass reframed Bowen's dilemma by suggesting that calling his efforts and explorations "research" could give them credibility.

The Menninger Foundation had a Research Department. Bowen had not formally tried to pursue his interests there. But in July 1953, after Menninger's presentation, Bowen was considering how to advance his own theoretical explorations. A staff discussion by a social worker about a woman with an alcohol problem had caught Bowen's interest. Bowen made one last effort to move forward within his present workplace. The exact date of Bowen's conversation with Kass is unknown, but it prompted his submission of a proposal.

At a staff conference in April 1953, Bowen heard a report on a 31-year-old divorced woman whose mother had previously used deception to transfer her alcoholic daughter to another hospital out of state without informing the daughter. She had now been at Menninger for two years (Bowen, 1953c, 1). Bowen recognized this patient might be a candidate for a sanctioned project using his regression treatment, though she was now under the care of another psychotherapist at Menninger. Bowen began meeting with the patient by June when her primary doctor was leaving Menninger's employ (Bowen, 1953d, 2). By July when the patient had discontinued formal psychotherapy, Bowen became her Hospital Doctor and psychotherapist out of his interest "in working with such a patient" (Bowen, 1953e, 1) and selected her for a formal research study. She was by then deemed a treatment failure, and her mother had interfered in her treatment, prime findings of interest to Bowen (Bowen, 1953d, 2). His clinical notes are insightful.

> She (the patient) described in great detail how she could not bear to be away from her mother when she was away, and how she didn't feel she could stand to be with her mother when she was in the house with her.
>
> *Bowen, 1953f*

After working with her for two months, in August Bowen submitted a proposal on alcoholism to the Research Department with her as the subject. This was his last endeavor to act on his ideas at Menninger. The proposal had many of the same aspects previously used with hospitalized alcoholic and schizophrenic patients and with out-patients. He based the plan on psychoanalytic principles, suggesting Menninger was unique in offering continuous transitioning from in-patient care to day program care (Bowen, 1953d, 2). Menninger would incur no cost. As the patient had a period of "good standard treatment, she would … not be deprived of a result by an experiment with less orthodox methods" (3).

For the proposal, Bowen addressed the parent–child transference relationship in the case record. He described the basis of the regression efforts and treatment to be explored, especially staff making no demands that could interrupt the cycle of regression and growth.

People with such severe passive oral character formations go through life with one outstanding motivation, and that is to maintain a state of homeostasis with the mother figure. It is as if they must maintain this umbilical cord-like relationship with the mother. The length to which they can 'stretch' the cord depends on the attitude of the mother figure ... the mother can never meet all of their expectations at which time they will either get drunk or go into some other kind of collapse .... If this mother figures makes demands (and the mother figure can either be a real mother, wife, husband, therapist or other supportive figure), they become terribly anxious ... the continual pattern that repeats itself over and over is demand—anxiety—compliant attempt to meet the demand, failure, collapse, return to mother to be accepted and babied or kicked out.

Bowen, 1952, 1

He worded his research proposal to advance Freudian theory. It "... would be in the framework of psychoanalytic principles... to achieve a high degree of integration between psychoanalysis and the milieu treatment program in a psychoanalytically oriented institution" (Bowen, 1953d, 2). The hypothesis stated that very early interactions between parent and child influenced the later emergence of alcohol dependence. This proposal intended to examine a way to increase understanding of parent–child relationships and offer a direction for growth out of alcoholism. Psychiatry considered such a possibility nearly impossible with alcoholics. Bowen proposed that a transference state could occur given specific milieu conditions and special use of nursing involvement (Bowen, 1952, 3). He had already achieved this in his previous investigations. The key was to develop a transference that replicated the original family relationship. This transference occurred in psychotherapy or psychoanalysis with mediating supports from a milieu that limited the number of people involved and maintained consistency in the staff providing care. There would be two people taking part, the Hospital Doctor/therapist (one person) and a nurse (4). These two staff were crucial to the plan's potential success. Basically, Bowen intended to recreate the family constellation and approximate a familiar yet unconstrained environment of relationships. The total possible relationships would be limited to parental surrogates to have fewer variables to track.

If accepted, Bowen's project would have Menninger's endorsement as research. Bowen's earlier explorations, covered in the last chapter, did not have this status. The decision to study alcoholism was a nod to the real world. A symptom spurned by society, chronic, severe alcoholism, was seen as a legal problem and alcoholics were jailed. Treatment facilities commonly did not accept alcohol or drug addicts. They were designated as "... a class of psychiatric 'untouchables'" (1). As Bowen wrote in the proposal, "the oral addictive character structure is '... psychologically understandable and psychologically reversible'" (3). Being sanctioned by Menninger could bring positive recognition to the Menninger Foundation and to Bowen's efforts. Menninger offered full-service supportive milieu programs that could allow for comprehensive study of the problem of alcoholism and an effective treatment.

The general outline for this research followed Bowen's regression explorations and added what Bowen had learned from the Shawnee Guidance Center experiments. That implied he would see the woman's family as well. Bowen's research would question psychological factors, not constitutional or organic contributors (3). The plan incorporated aspects of support for the patient's individuality and the taking of responsibility for self. In Bowen's own words, he described the proposal as "a more systematic and controlled study" than his earlier experiments (4). He would replicate the parents' function in interactions with the patient that produced a transference while also involving the family.

> It is my hypothesis that a very intense and completely workable transference can be built on actively 'doing' for the patient, and that the patient's perception of the doctor in terms of the original parental figures is not sufficiently impaired by the 'reality' of the situation to prevent an excellent and workable transference.
>
> 7

"Doing" meant giving freely by the central participants: nurses, aides and clinicians in response to patient's direct requests. Another definition of "doing" was "motherly attention" (5) accepted by the patient.

> At this point, when as much 'full time mothering attention as the patient could actively request' is in place the transference develops rapidly and very soon the patient finds himself completely enmeshed in an intense oral dependence in which is acted out the original triangle between father-mother-child.
>
> 6

The "triangle" was the point of potential growth. If the patient's acting out had not forced the staff operating as family to withdraw or had not pulled the staff in to take over for the patient, then the slow resolution of the transference toward more maturity would take place. The staff would need to recognize their own behavior. The turning point would come when the patient acted to end the staff's active "doing." Interestingly, Bowen noted that this approach is "too laborious and difficult and time consuming" for general usage. He then noted that the same holds true for psychoanalysis, "which is too expensive and time consuming for general treatment" (7). He described his greatest contribution from his explorations into theory and treatment as "the understanding of people" or an effort toward making sense of human life. When the innate growth kicked in, Bowen had a way to be present that slowly resolved the transference. It balanced separateness and closeness. Could a future program teach parents these methods as he had taught them to the medical residents?

Bowen actively worked to make this a groundbreaking project. He invited the committee to think through the project with him and offered to provide an open forum for discussion of observations over the course of the research. That forum could then serve as a platform to extend the ideas. The research members would have a role beyond approval and a progress check. Bowen asked them to function as a brain trust, an appeal to include the Research Committee in innovative explorations on the science of human nature that could attract other professional research fellows interested in studying the human. The proposal offered a solution to the problem Bowen identified in the July memo to Karl Menninger. Acceptance of broad new ideas is greater when it has supporting research. His ideas were so revolutionary that without scientific corroboration, it was likely people could easily overlook them. Bowen was intrigued by Menninger's broad theoretical assumptions and offered a project to explore them.

Bowen held the Menninger complex of services in high esteem. Menninger administration reciprocated and valued Bowen. His clinical skill was highly respected, yet his theoretical ideas were on the fringe. The Committee declined Bowen's proposal. There is a balance of independence and interdependence that sets the emotional climate in a family raising a child. This balance determines the future course of that child's life. A work setting similarly provides an emotional climate for its employees. Bowen had positive regard for Menninger, but he set his goal of contributing to a scientific theory of the human above his positive relationships. The Menninger Research Department missed an opportunity; Bowen moved on.

In my research, I have wondered about this missed opportunity and how to understand it. Refusing the proposal might have been alcoholism bias. Or because Bowen had informed the administration at the beginning of the year that he might leave. Was it Bowen's subtle attempt to connect Freud with evolution? Replicating a family's transference contradicted established practices. Altering a person's trajectory by having the offspring work out transference in relationships with the parents was beyond the committee's grasp. Bowen's effort to communicate his early idea of the family as a unit had failed. Yet, his proposal is a document of Bowen's intellectual understanding then in place. It supports his claim that the theory was formed, with a hypothesis to explore it, while at Menninger.

Perhaps the refusal came out of already reviewing a project that was more wide-reaching in terms of the number of participants and researchers or it was just bad timing for Bowen to submit a proposal. Parallel to Bowen's proposing his research project, Robbins and Wallerstein were organizing a longitudinal project, "The Psychotherapy Research Project." Lewis Robbins was a Menninger resident in the early 1940s before the school of psychiatry opened. After the war, he returned there as Director of the Outpatient Division (1946–1949) and then as the Director of Adult Psychiatry through the mid-1950s (Menninger, 1984). Robert S. Wallerstein, another resident at Menninger (1949–1951), later became Director of Training and then Director of Research. Both would have been familiar with Bowen and his clinical work. The Robbins/Wallerstein project would run at Menninger from 1954 to 1982.[1] That study followed 100 patients and compared treatment processes and outcomes (Wallerstein, 2002). Half the subjects received psychoanalysis. The other half received psychoanalytic supportive psychotherapies. Neither psychoanalyst nor therapist had knowledge of who were the selected subjects (Galatzer-Levy,

Bachrach, Skolnikoff, & Waldron, 2000, 72). This research project was internationally known.

> The influence of patient, analyst and environmental factors on clinical outcomes were studied, along with the basic postulate of the psychoanalytic theory of therapy. New methods were developed, and scores of investigators, as well as distinguished consultants from around the world, were enlisted in the project. In its thirty-year history, it produced more than sixty papers and five books.
>
> *63*

In 2000, Galatzer-Levy, Bachrach, Skolnikoff and Waldron, all psychoanalysts, evaluated the efficacy of psychoanalytic treatment in major studies including the Robbins and Wallerstein project. They reported a surprising finding that using an approach that strengthened a patient's innate adaptive abilities, was superior to fostering insight and relying on an analyst's interpretation (68–69). That was the core of Bowen's proposal in 1953. He had integrated his new thoughts on a human emotional system and differentiation of self. The method in his proposal would redirect the transference from staff back to family. It was ahead of its time.

Robbins sent a memo to Bowen on August 13, 1953, noting that they did not approve the project proposal as well as his inability to understand Bowen's operating theory (Menninger Research Committee, 1953). Perhaps it was that the Wallerstein and Robbins project stayed within accepted standards of practice. Bowen reflected later that the deleterious effect of "… orthodox psychoanalytic thinking may have delayed the development of the family movement" (Bowen, 1979, 2).

Bowen's proposal, if accepted, could have set a precedent, leading the investigation of alcoholism and understanding it as a treatable malady, not as a crime. But he was not one to linger in fruitless activity. Orthodoxy was an impediment to fact gathering. When the committee refused his proposal, it settled the question. No longer was leaving Menninger an intellectual exercise. Bowen immediately began a national search for another position.

A brief case discussion

So what happened with this woman intended for Bowen's research project? On a visit home in early August 1953, she engaged the family in assisting her to get better. This prompted her mother to phone Bowen and her sister to visit him in person (Bowen, 1953e, 1–10). Bowen described the relationship between the two sisters as "intense and hostile," based on the sister's perception that mother's attention to the patient deprived her (the sister) of a mother (Bowen, 1953f, 2). Bowen was interested in how hospitalized people could appear to be functioning well but on return to their family, lose the gains made in hospital. The patient had a "suicidal episode" two days after her home visit and one day after mother and sister spoke with Bowen. In the records available, Bowen continued psychotherapy with her into the fall. He may have continued longer, but the record trail ends here. He used another version of paraldehyde and the non-mothering approach with her on the ward. She took on more responsibility for her own use of paraldehyde and for her relationships with her family and was functioning more independently.

The records show the intensity of the family relationships. The way Bowen helped the woman by relating to her mature side or, as written in the clinical narrative, to the best of this patient (not the inner child) (3) is another example of his practice of being present, of being a resource while staying inside his own skin. Bowen later used this method, of believing that adult-oriented thought exists among the patient's infantile communications, listening for it and responding to that thought, leaving the infantile side in the family and interacting with the strength in the person, with staff on the National Institute of Mental Health (NIMH) project.

Timely opening of the NIMH Clinical Center

In April 1953, the federal government made health care a cabinet-level priority by transferring the Public Health Service (PHS) to the new Department of Health, Education and Welfare (The NIH Almanac, 2016). The department planned to open a Clinical Center in Bethesda, MD. A 14-story hospital with 540 research beds solely dedicated to clinical research, surrounded by twice that number of scientific laboratories (US Department of Health and Human Services, 1980, 8). Having the national government conduct clinical studies was a change. Formerly, mainly mice and primates were studied; now human subjects would be researched. Creative thinkers envisioned a scientist going from the lab to clinical contact and back to the lab, while interacting with other scientists on the way, to and from. The aim was to create a self-contained assemblage of clinicians, scientists, patients and support staff, with the common goal of conquering both chronic and acute disease.

The Clinical Center at the NIMH promised a boundless opportunity for experimentation and original ideas pioneering progress in studying body and mind while supplying a home for pure research that extended the clinical dimensions of the existing PHS programs (NIH Clinical Center, 2018). Bowen wanted a setting that gave him freedom to pursue his revolutionary idea of leaving transference, the infantile side, in its developmental relationships while the therapist became an active resource to patient and family. The family as a natural unit was not yet a cohesive theory put into words.

Reasoning and observable facts were his tools to find a way toward general principles. He needed a place to create conditions that would generate insights into parent/child interdependence, to clinically study these conditions, and to look for exceptions.

Bowen had confidence that the concepts originating in his investigations could replace the soft concepts in Freudian theory. He was looking for an environment that offered freedom of thought and stretched current belief systems, where extending current knowledge was encouraged (Bowen, 1986, 5). An inquiry at the University of Wisconsin, where he had presented in 1951, brought a quick reply in

September offering a professorship, with the ability to found and chair a department (Bowen, 1953g). Bowen wanted to split his time between research and private practice. He wanted a place where he could generate hypotheses on parent/child interdependence, to clinically study these, learn and look for exceptions. This position wasn't quite the fit he wanted.

Relationship connections prepared the way

Bowen presented a paper titled "Psychotherapy in a Guidance Center" in September at the Annual Clinical Meeting of the Mid-Continent Psychiatric Association, in Kansas City. This paper prepared Bowen for the interview offers he was soon to receive and was a rehearsal to present his work effectively.

In the presentation, he spoke of his shift from healing people to aiding their own efforts to resolve their own problems (Bowen, 1953h, 2). Doctor and patient each had defined responsibilities. The doctor was to be present to the patient in all ways possible and the patients were to make use of that resource to advance their own growth. He reviewed how the Guidance Center program operated. There, the resident resisted the impulse to be the expert. And the patient accepted an active part in solving his or her own personal dilemmas (3). The emphasis was on the patient's strengths and on increasing the level of maturity. The main body of the presentation was on the application of this method to the diagnosis of schizophrenia. Bowen noted that the patient always asked two questions in this work: "What do you think of me?" (10) and "What do you want of me" (11)? Bowen specified that answering as one adult to another was more effective than responding as if it were the bid of a child to a mother. The non-mothering approach maintained the "neutral, supporting, non-directive attitude" (17). He reviewed the difficulties of using this method. The residents who had the "capacity to trust the strength" of even seriously impaired patients did the best (9). An interesting note was the increased effectiveness of residents when implementing this approach (6).

Bowen traveled cross-country searching for a position that encouraged innovative studies. He interviewed at NIMH in November 1953 (Bowen, 1977, 4) and at Yale University, in New Haven, CT (Bowen, 1953i). He visited Duke University (Bowen, 1954a) and sent inquiries to the Washington Institute of Mental Hygiene (Bowen, 1953j) and Northwestern University Medical School (Bowen, 1954b).

> I began to look around the country for places that I could do … half-time research and half-time therapy. I investigated a number of places and wrote letters and it was sort of a fortuitous thing that got NIMH into it.
>
> *Bowen, 1953a*

That "fortuitous thing" was the staff recruitment processes at NIMH's Clinical Center. NIMH needed staff with innovative research ideas who were willing to work near Washington, DC. The clinical center was given a mandate to gather new knowledge with a promise that might lead to disease prevention. Reversing chronic alcoholism or possibly schizophrenia was in synchrony with that mandate.

Bowen's interview at NIMH and the road to Bethesda that followed in 1954 had a backstory involving relationships going back to at least 1946. Robert H. Dysinger (who, from 1954 to 1959, became a co-investigator on Bowen's NIMH project) had accepted a position at NIMH as assistant administrator and research psychiatrist in July 1952 (Dysinger, 1989). It was in December 1952 that Robert Cohen, the man who would hire Bowen at NIMH, reported to his position at the not yet opened NIMH (Farreras, Hannaway, & Harden, 2004, 186). Director Cohen asked Dysinger to see Louis Cholden, as a potential research candidate for the Clinical Center. Cholden, a resident who Bowen supervised at the Menninger Clinic, was recruited to work at NIMH by its director, Robert Cohen, before NIMH had even opened. Cholden got a "thumbs up" from Dysinger and he was hired (Dysinger, 1989, 2). Once Cholden felt established there, he commented, "I know someone who could really do something with this place" (Bowen, 1972, 2). Meanwhile, Bowen,

> in talking to one of the psychologists at Topeka, who w[as] saying that Lou Cholden was at NIMH ….The clinical center had just been open then a few months ….That night I wrote Lou a letter and two days later I had a telephone call from him urging me to come to NIMH.
>
> *Bowen, 1972*

Within a week, Bowen wrote to Cohen expressing his interest and that he had "a yearning to really get into [certain problems] and see them through" (Bowen, 1953k, 1). Dysinger claimed indirect credit for Bowen coming to NIMH because he had hired Cholden. Coincidentally, Bowen's interview with Cohen at NIMH in November overlapped with the opening of Ward 3 East at the Clinical Center and admittance of the first adult schizophrenic patients (Bowen, 1969; Farreras et al., 2004, 71). This was the unit where Bowen would establish his study project the next year.

The Bowen family stayed with the Cholden family when Bowen came for the NIMH interview. "My former student became my entrée to the world of psychiatric research" (Bowen, 1968, 2). Having fulfilled military service years ago, Bowen was an attractive candidate. Cohen's suggestion that they could be bringing together a top-notch research organization had a powerful impact on Bowen in the interview. "Of all the places I had gone I compared them to NIMH, and I went back home and went about the business of putting in an application for the following July" (Bowen, 1972, 2).

Correspondence between Bowen and Cohen continued into the New Year. They addressed several points including Bowen continuing with more analysis for himself (Bowen, 1954c, 1) and arranging for any patients who might transfer with him, including one analytic control case (2). Cohen and Bowen both agreed to this. In January 1954, Bowen accepted a position at NIMH (Bowen, 1966). The position was full-time research psychiatrist at NIMH with time for private practice.

Bowen left Menninger's on June 12, 1954 (Bowen, 1954d, 1). By the time Bowen arrived, in July 1954, Dysinger had left his administrative position and was the working professional in a study involving "major psychotherapy services" (Dysinger, 1989, 2). Ironically, Bowen became the supervisor to Dysinger's group (Bowen, 1954d, 2).

Reflections on the Menninger years

Bowen discussed theoretical possibilities during his years at Menninger, but the timing of an alternative model based on facts was not right.

> Thoughts about a different theory were beyond popular comprehension when Freudian theory was still in ascendancy. The ideas [Bowen's] were never secret, but they were never advertised. It was a 'toe dance' that also involved the difference between feelings and facts.
>
> *Kerr & Bowen, 1988, 358*

His theory on human behavior was, in his words, "a radical departure from traditional beliefs" (Interview with Murray Bowen; Bowen's behavior idea has rocked Psychiatry, 1977). "It is so far out of the established that the majority of the profession would not accept it …. If you think it is far out now, you should have been around in the 1940s when I originated the theory" (Bowen, 1977, 1). Even by 1971, a resident then at Menninger referred to Bowen's idea, noting "That was really pretty far out, the idea of doing something with the whole family was really pretty far out" (Beal, 2018).

It was a complex decision to leave the Menninger Foundation. Faced with taking a pay cut to start somewhere else, the future was uncertain. Bowen was a middle aged, married man with four young children. His children had been born in Topeka and the family was established in the community. Relationships there went back to his Army days. The environment at the Menninger Clinic gave him latitude to pursue early insights that were stimulated and refined by his experiences first as a resident and then as a staff member. Years after leaving Menninger, he expressed his gratitude for the magnitude of his growth during those years and for his mentor, Karl Menninger:

> A systems way of thinking has a way of conceptualizing the disparate paradigms of psychology, biology and sociology. Serious psychiatric problems involve genetics and physiology and psychology and sociology. It may take another one hundred years but I believe that what systems will eventually show the way to those who try to hang onto the previous way of thinking, I, and those who follow me …, will keep on with this as long as it takes. I will remember that you are one of the older people with an open mind. The association with you from 1946 to 1954 probably had more influence on me than

all the years spent with my biologically oriented father who knew we were ONE with all the animal and plant life that surrounded us in rural Tennessee.

Bowen, 1981

Bowen had understood that the more recognized an institution is, paradoxically, the greater the expectation that the institution preserves the standard approach (Bowen, 1969, 2). While Bowen theory attracts trainees from around the globe and was (and is) taught worldwide in universities "… The Menninger mecca was one of the last places to change. By 1985 the new theory had pervaded Topeka. The Foundation invited me back to receive a special award, and to claim me as its own …" (Bowen, 1987, 1).

Bowen wrote in his last decade of life that he came to Menninger with an "orientation in the science and art of medicine" and a "conviction that the human mind could be as much of a science as the rest of the human" (Kerr & Bowen, 1988, 347). Bowen's time at Menninger served as the crucial, foundational years for the creation of a new theory. Immersing himself in psychiatry and Freudian theory, Bowen arrived at the unexpected goal of making Freudian theory more scientific. In the new Menninger School of Psychiatry, innovative in its application of Freudian theory and progressive in its holistic understanding of human life, Bowen emerged from the residency program with pioneering ideas for both the theory and its practical applications. One tenet at a time, he methodically appraised Freudian doctrine, gathering facts to replace the assumptions. Bowen's new primitive theory differed in both large and small ways from Freud's. Bowen considered Freud a great man, who oriented human behavior toward understanding a primary human relationship, mother and child. But the focus on detecting the facts of human behavior connected to a family system separated Bowen from Freud. Bowen had given up the primacy of the individual.

When Bowen's time at Menninger's ended, a new and productive era began. Research interest and relationship connections brought Bowen to NIMH. They began in Bowen's Army days with Norman Brill and Will and Karl Menninger, continuing through his residency at Menninger's, and later as staff supervisor to Cholden. Cohen, charged with staffing this new research facility, had a vision of unfettered exploration, astute foresight in selecting staff, and he had principles on operating a research setting that were a good match for Bowen's interests. A new theory was the result.

Note

1 For a review of this and other systematic studies of the efficacy of psychoanalysis see *Does Psychoanalysis Work?*, Galatzer-Levy, R., Bachrach, H., Skolnikoff, A., & Waldron, S. Yale University Press: New Haven, 2000.

References*

*Unless otherwise specified, the works of Murray Bowen included in this chapter are from the Murray Bowen Papers. 1951-2004. Located in: Modern Manuscripts Collection, History of Medicine Division, National Library of Medicine, Bethesda, MD. The Accession, Box and Folder information differs for each one and that is included here.

Beal, E. (2018). [Oral history interview, May 2018, A. M. Schara, interviewer]. Retrieved July 30, 2020, from The Murray Bowen Archives Project: Oral History Interviews, http://murraybowenarchives.org/oral-history-17-Beal.html.

Bowen, M. (1952). [Menninger clinical record, conference summary, November 21, 1952]. Copy in possession of Bowen family, Williamsburg, VA.

Bowen, M. (1953a). [Letter on January 26, 1953]. Copy in possession of Bowen family, Williamsburg, VA.

Bowen, M. (1953b). [Memo from Bowen to Dr. Karl, July 8, 1953]. Copy in possession of Bowen family, Williamsburg, VA.

Bowen, M. (1953c). [Menninger clinical record, conference summary, April 21, 1953]. Copy in possession of Bowen family, Williamsburg, VA.

Bowen, M. (1953d). [Memo to research committee on August 8 re: Alcohol research proposal]. Copy in possession of Bowen family, Williamsburg, VA.

Bowen, M. (1953e). [Menninger clinical record, name redacted, July 1 to August 12]. Copy in possession of Bowen family, Williamsburg, VA.

Bowen, M. (1953f). [Menninger clinical record, August 13, 1953 to August 25, 1953]. Copy in possession of Bowen family, Williamsburg, VA.

Bowen, M. (1953g). [Letter on September 10, 1953]. Copy in possession of Bowen family, Williamsburg, VA.

Bowen, M. (1953h). [Paper. Psychotherapy in a guidance center, September 26, 1953]. Copy in possession of Bowen family, Williamsburg, VA.

Bowen, M. (1953i). [Letter to E. S. Kessler on November 12, 1953]. Copy in possession of Bowen family, Williamsburg, VA.

Bowen, M. (1953j). [Letter to Frederick C. Redlich on November 13, 1953]. Copy in possession of Bowen family, Williamsburg, VA.

Bowen, M. (1953k). [Letter to Robert A. Cohen on November 11, 1953]. Copy in possession of Bowen family, Williamsburg, VA.

Bowen, M. (1954a). [Letter to Ewald W. Busse on January 6, 1954]. Copy in possession of Bowen family, Williamsburg, VA.

Bowen, M. (1954b). [Letter to Benjamin Boshes on March 20, 1954]. Copy in possession of Bowen family, Williamsburg, VA.

Bowen, M. (1954c). [Letter to Robert A. Cohen, M.D. on January 6, 1954]. Copy in possession of Bowen family, Williamsburg, VA.

Bowen, M. (1954d). [Letter to a patient's husband on July 10, 1954]. (Acc. 2007-073, Box 3, Folder Correspondence/Patient).

Bowen, M. (1956). [Letter in response to an inquiry from South Africa about training opportunities at the Clinical Center on September 6, 1956]. (Acc. 2006-003, Box 6, NIMH correspondence a through G.).

Bowen, M. (1957). [Letter to a colleague on February 20, 1957]. (Acc. 2006-003, Box 6, Folder NIMH correspondence H through P).

Bowen, M. (1966). [Explanation of IDENTIFYING DATA section]. (Acc. 2009-013, Box 6, Folder GAP papers).

Bowen, M. (1968). [Letter on December 23, 1968]. Copy in possession of the Bowen family.

Bowen, M. (1969). [Letter on March 16, 1969]. Copy in possession of Bowen family, Williamsburg, VA.

Bowen, M. (1972). [APA interview with Mark Johnson on February 2, 1972]. (Acc. 2007-012, Box 2, Folder APA interview: Dr. Bowen and Dr. Johnson).

Bowen, M. (1977). [Interview with Murray Bowen; Bowen's behavior idea has rocked psychiatry]. Clipping from News-Democrat, Waverly, TN, 28:36, 1–2, September 7, 1977. Copy in possession of Bowen family, Williamsburg, VA.

Bowen, M. (1981). [Letter to Karl A. Menninger on July 6, 1981]. (Acc. 2007-012, Box 3, Folder Correspondence between Dr. Karl Menninger and Dr. Murray Bowen).

Bowen, M. (1986). [Presentation. Working with the extended family: The therapist and the therapy, June 7, 1986]. (Acc. 2003-044, Box 6, Folder CFC – Chicago).

Bowen, M. (1987). [Letter to a former client on July 31, 1987]. (2007-012, Box 2, Folder Re: Deaths and funerals).

Bowen, M. (1979). Part II—Adventures in 'the great democracy': An interview with Murray Bowen. *Family Therapy Practice Network Newsletter, III*(5), 1–7.

Dysinger, R. (1989). [Letter to a PhD candidate on September 11, 1989]. (Acc. 2007-012, Box 4, Folder Bob Dysinger).

Farreras, I., Hannaway, C., & Harden, V. (Eds.). (2004). *Mind, brain, body and behavior: Foundations of neuroscience and behavioral research at the National Institutes of Health, Washington, DC.* Amsterdam: IOS Press.

Galatzer-Levy, R., Bachrach, H., Skolnikoff, A., & Waldron, S. (2000). *Does psychoanalysis work?* New Haven & London: Yale University Press.

Kerr, M., & Bowen, M. (1988). *Family evaluation: An approach based on Bowen theory.* New York: W.W. Norton.

Menninger Research Committee. (1953). [Memo re: Doctor Bowen's research treatment proposal, August 13, 1953]. Copy in possession of Bowen family, Williamsburg, VA.

Menninger, K. (1984). In Memoriam: Lewis L. Robbins, M. D. (1913-1984). *Bulletin of the Menninger Clinic, 48*(5), 455. Retrieved February 10, 2021, from https://investigation.proquest.com/openview/fd900e42ada13889818a46080a96cbdf/1?pq-origsite=gscholar&cbl=1818298.

NIH Clinical Center. (2018, January 12). Retrieved July 3, 2020, from Dept of Health and Human Services, National Institutes of Health: https://cc.nih.gov/ocmr/history/timeline/index.html.

The NIH Almanac. (2016, October 27). Retrieved July 3, 2020, from National Institutes of Health: https://clinicalcenter.nih.gov/ocmr/history/index.html.

US Department of Health and Human Services. (1980, June). The National Institutes of Health. NIH publication no. 80-1.

Wallerstein, R. S. (2002). *Psychoanalytic therapy research.* Retrieved February 10, 2021, from Psychomedia: www.psychomedia.it/spr-it/artdoc/waller02.htm.

10

SETTLING IN AT NIMH

July to October 1954

First, it is insightful to review the "Annual Report of the Bowen Section" (1952–1953) from Menninger. Produced a year and a half before Bowen began his research at the National Institute of Mental Health (NIMH), the section report foreshadowed Bowen's beginning effort at Bethesda, where he extended his principles of an open ward through, involvement of every staff member, attention to staff's functioning, both their strengths and weaknesses, routinely reviewing what was being learned and what needed more scrutiny, and operating the project in the broader environment. The interesting points in this report discussed personnel, milieu and Bowen's recognition of the process in the group. Bowen addressed when it worked and when it didn't. Stability in the staff offered strength and cohesion in the unit's operations and goals.

Bowen oversaw two junior psychiatrists, a resident psychiatrist, a psychologist and two psychology interns, plus a social worker, a nurse, an adjunctive therapist, marriage counselor trainees and a secretary. All attended the research meetings. The secretary's attendance at section meetings, deemed important to the gestalt, represented a nod to an open group.

This unit had three section meetings each week because the work volume was so great, although each meeting also added eight to ten hours a week to that volume. Yet the report showed that work volume went better with three meetings in order to give staff time to discuss therapeutic methods, patient progress and the expectations of the larger hospital. Attention went to staff–staff interaction, staff–family interaction, family–family interaction and ward–ward experiences.

In the report, Bowen identified each staff member by name, acknowledging their individual progression to better functioning and improvement. As to their functioning as a group, Bowen noted areas still in need of progress. Bowen's section carried the highest patient load of all sections, with 63 total patients including readmissions (Bowen, 1953, 3–6). He left staff to decide for themselves what they

DOI: 10.4324/9781003027287-11

could do, expressing his approach as "… the positive philosophy 'of choice in the kind of a load they are capable of carrying rather than urging them to meet a quota set by their supervisors.' The review noted that this "positive philosophy lost its efficiency" when competitive tensions, comparable to those between siblings, arose between two of the professional disciplines involved in the section.

This report addressed the section's intragroup pressures, intergroup strains and positioning within the broader hospital, revealing the inner workings of Bowen's section. As for the staff doctors, Bowen restated his position that the individual doctor was the better judge of his or her own workload competency. (This was the same approach Bowen used with patients, letting them find their way while he stayed present and close by. Personal initiative was most important. Don't interfere with that.) Staff observed themselves as well as the patient checking their own acceptance of personal responsibility. It was an environmental approach, looking within the unit and within the larger organization.

Bowen's unit had a liberal visitor policy, allowing visitors to attend these section meetings. A comment from a visitor who attended a meeting complimented the meeting's progression.

> After the meeting, he said he enjoyed the meeting very much, especially the social phenomenon he had just witnessed. He said the Section Leader had expressed a viewpoint early in the meeting and that following this every member of the group except one had disagreed with the leader's viewpoint and had gone on to express his own viewpoint, following which the entire group had worked out a satisfactory solution. He said the social worker was the only one who had not disagreed with the leader and that she took a mid-point stand.
>
> *Bowen, 1953, 5*

The report also provides a window into the influence of Bowen's learnings from his out-patient efforts on in-patient operations, specifically the positive effects of increasing the staff's and patients' personal responsibilities. At NIMH, where the parent and the daughter were hospitalized together, Bowen applied this learning by making a major modification to the nursing staff role and leaving the caretaking between the two family members.

The Clinical Center at NIMH: A very different research environment

Exactly one year after the Clinical Center opened, Bowen arrived at NIMH. July 1 is traditionally the date when new residents and new programs begin in psychiatric hospitals. One psychiatric ward was operating then with locked doors, and there were 11 patients. The Center's emphasis on inquiry allowed for smaller patient numbers than did other research places that were focused on outcomes. Bowen remarked on the uniqueness of the patient-staff ratio at NIMH.

This all does not make sense according to ordinary standards, but an example from the Cancer Institute, which is also in the same building, will give you an idea. There is an entire staff of doctors, nurses, technicians, and others all devoting their entire time to about 4 patients.

Bowen, 1954a

The Clinical Center, in July 1954, was a bustling, expanding facility. Bowen's arrival was part of that momentum. Bowen expected two of his Menninger regression patients, one with severe schizophrenia and one with acute alcoholism, to follow him to NIMH. He arranged with the administration there for this to happen.

The extra staff coming on board freed Cohen's attention from recruiting to managing clinical efforts in psychotherapeutic programs. As the Director of Clinical Investigations, his vision for NIMH was now coming to fruition (Cohen & Jenkins, 1954). Cohen appointed Bowen as Director of Psychotherapy. This was unforeseen, and the position came with responsibilities in getting that program up and running and in providing hours of supervision (Bowen, 1954b, 1). In a letter to a patient's husband from Menninger, ten days after arrival, Bowen opens a window into the unique context of researchers operating in an open-ended climate.

I find myself the Director of the Psychotherapy Division. Dr. Cohen is Director of the entire operation and he has set up three divisions .... The third is my Psychotherapy group. I have no very clear-cut ideas about how I will proceed but we'll get something going .... Autonomy here is complete. No half-way measures .... This place comes as near to my concept of a psychiatric utopia as can be devised. Most important is the attitude of the people .... The practical mechanism for working is that the doctor has a particular kind of a problem on which he proposes to do research .... First he gets his project lined up and the people who will help him. He also has to get an okay from the group to use a bed. Then he finds the patient to fit his project ....When the doctor finds the patient, he then brings the patient onto this rather ... ideal setting at absolutely no cost to the patient .... Here the patient is a patient of a particular doctor and the place is here to help the doctor in what he wants to do ... the speed with which I get such a project under way will depend on how rapidly we can get this whole thing integrated and know where we stand in relation to each other. It might be in October of this year. It might be spring. It might be a year from now ... everything has started quite well for me here. People like my ideas and my viewpoints.

Bowen, 1954b, 1–2

In September, Bowen acknowledged his early exuberance had been a coping mechanism to manage the emotional response of leaving Menninger (Bowen, 1954c, 1).

Bowen's caseload in the first year shows that he had little time for anything but work. On arrival in Bethesda, he established a private practice, finding an office in a building owned by a colleague. Bowen's appointment book was full within four days of opening his practice (Bowen, 1954a, 1).

Early on, the head of the NIMH called its new group of researchers together and asked each one "Why are you here?" Bowen replied he had one answer to schizophrenia and he wanted to give this answer a run (Bowen, 1985). Privately, he reflected on his hubris.

> If I knew so much about what caused schiz[ophrenia], [I] Should be able to take a newborn baby and create schiz[ophrenia] …. No way to do such an outlandish lifetime project, but I dulled my head on that basic proposition. 'If we really know what it is we are treating, we should be able to create it'. Early in research, I had moved pretty much into <u>natural systems ideas</u>.
>
> *Bowen, 1985; underline in original*

He was well on his way to what Westley, Zimmerman and Patton, who explored social change within complex systems, in capturing this gestalt and referencing other social innovators' effort to maintain perspective, as "simultaneously keep[ing] head in the stars and feet on earth" (Westley, Zimmerman, & Patton, 2006, 167). His early efforts at the Menninger Foundation to move Freudian theory toward science were now repurposed to move his new theory toward the goal of a broad base of facts.

The significance of one pre-research family, July 1954 to September 1955

Soon after his arrival, Bowen met with an out-patient family, beginning on July 22, 1954, whose presenting problem was with alcohol. I have named this family the Dutch family. Bowen saw them before and during his mother–daughter project. The evidence shows in the Dutch family's clinical record that Bowen's thoughts were crystalizing around the theory of family as a unit prior to his beginning the official NIMH project studying mother–daughter pairs. Sessions continued with the Dutch family through discharge in September 1955.

What follows comes from the clinical record. The family comprised a father, mother, eight-year-old daughter and an involved paternal uncle. The husband made the original contact, complaining that his wife was drinking a fifth of whiskey a day (Bowen, 1954a, 1). Bowen's assessment was that the underlying symptom was schizophrenia. The wife's alcohol use occurred when other mechanisms for getting on with living were not sufficient. The patterns later seen in the mother–daughter relationships, he first saw here. Bowen wrote about these patterns presumably without being fully conscious of a family unit perspective but the patterns are congruent with his later observations of the research families in his NIMH project.

It was Mr. Dutch's brother, not the wife or the husband, who showed up at the first appointment (Bowen, 1954a, 1). When Bowen saw Mrs. Dutch, which was before his in-patient schizophrenia project had begun, she came to the office with her husband and daughter. The next week, the mother and child came to the appointment. The child clung to her mother's leg as the mother attempted to go into the interview. Bowen left the decision on what to do about this to the mother, who then allowed the young girl to sit in on their session (9). By mid-August, there was no mention of drinking, but the marital relationship was deteriorating.

Both parents were totally focused on the daughter, each with increasingly different views about how to approach the daughter's problem. The wife now described the problem as being between her and the daughter (Bowen, 1954b). Neither parent had any idea about finding a resolution to problems in the marriage nor what to do about the child (Bowen, 1954c, 1). This couple was an example of parents presenting helplessness. If the therapist was willing, they would turn the problem over to him.

Bowen described the husband and wife as equally immature. Of note in these records was Bowen's countertransference with the father. Bowen sought consult with a peer to address this (Bowen, 1954a, 3). The mother improved, but the father began drinking. They then sent the child to live temporarily with the grandma (Bowen, 1955).

The Dutch family offers an excellent illustration of the shifting nature of family process when a family member does not consider self as part of family problems. Where was the problem? It was always in the other. Bowen's findings were a precursor to his observations during the first months of his in-patient research. In the clinical record, Bowen described a pattern in this couple of over- and under-functioning reciprocity (Bowen, 1954b). He saw the intense attachment between the husband and wife repeated between the mother and daughter. The couple's similar immaturity levels and shifts in functioning anticipated the reciprocity Bowen would see in the hospitalized mothers and daughters.

Here, with this couple and their child, Bowen observed upset moving among the threesome. Later, with the mother–daughter pairs, Bowen identified the shifting of functioning as a transfer of anxiety. Bowen wrote of this shift or transfer as "fluidity" of the fusion, noting the importance of timing. He wrote, "This mechanism was so common that any increase in mother's anxiety would alert the staff to expect an increase in the patient's psychosis" (Bowen, 1957a, 6). As one family member did well, the other did not.

In the Dutch family, which member did well shifted between the family members. When the marital pair was closer, awareness of the differences between them surfaced, and ironically, emotional distancing emerged to manage this new situation (Bowen, 1955, 5). Observing the husband and wife or the mother and child struggle with their emotional oneness led Bowen to replace the term symbiosis with the term oneness, at the end of 1955 (Bowen, 1959a). It is another example of the transformative progress that Bowen experienced working with them. This intact mother–father–child family showed the same patterns later seen in the in-patient mothers and daughters, lending credence to his gestating idea of a family as a deeply interdependent unit. With the mother–daughter pairs, the staff's difficulty of getting tangled in the relationship pointed to a missing third person. In this intact household, that third person, the husband, was present. Bowen saw the same sequences when he began hospitalizing entire families in 1956.

In the mid-1960s, while drafting a chapter on how Bowen arrived at his understanding of the family as naturally formed unit, he mentioned the importance of the data from less impaired families: "Concurrent observations of out-patient families suggested fathers were as involved as the mothers in the emotional oneness" (Bowen, c. 1965). Though not officially research subjects, the Dutch family pointed to the theoretical missing piece, a family as an interactive system. Out-patient work with the Dutch family, coupled with the in-patient observations, identified the patterns that were crucial to recognizing that all the members were a unit. That realization altered Bowen's hypothesis and led directly to his hospitalizing intact families. The scientist's comprehension was evolving with his recurring exposure to the facts.

The time of contact with both out- and in-patients overlapped. The Dutch family clinical records document Bowen's observations of the early emergent properties of the family unit in this family's interactions before his official research began (Bowen, 1959b). What he saw in the intact family mirrored the fragmenting process between the mothers and daughters involving the staff. When answering a request for medical records in 1960, Bowen had an epiphany. He had perceived the same patterns in the Dutch family that he was seeing with the in-patient families.

> I had a request … to send a summary to Mrs. [Dutch's] psychiatrist .… That was my first occasion to review the … record since 1955. I had worked with Mrs. [Dutch] in the period just before I started the family research and the … record was never a part of the active family study. In the course of preparing the summary, I spent quite a bit of time reviewing the [family] record. I was pleased to find that, in the light of our more recent experience in working with full families, I was able to see clinical patterns [in the records] that I had not seen when I worked with Mrs. [Dutch] … in 1954-1955.
>
> *Bowen, 1960*

Reading these clinical notes, it is striking how clearly Bowen understood and described this seminal finding before announcing it in March 1955. "Though the world does not change with a change of paradigm, the scientist afterward works in a different world" (Kuhn, 1970, 121).[1] This work with the out-patient family was the prelude to the in-patient discoveries.

Designing the research project August 1954

It was not a requirement that researchers at NIMH have a defined project. Bowen was clearly interested in developing a more scientific understanding and a general direction for proceeding. But he did not have a preconceived organizing structure for how to further his interests. He had joined NIMH with an open mind.

In a 1972 interview, Bowen said, "I didn't come with a definite idea of putting mothers and patients into the hospital" (Bowen, 1972a, 2). He wanted to investigate the parent/alcoholic offspring but could not get approval to have so-called drunks in the hospital. In fact, shaping a research effort around alcoholism proved problematic at both Menninger and NIMH.

> Back at the Menninger Clinic in the later 1940s and early 1950s, alcoholism and schizophrenia were my two main clinical interests. Those were the days when we still believed these problems could be 'analyzed' ....When I came to the National Institute of Mental Health to start my family research in 1954, I wanted to do the research on the intense relationship between the alcoholic and his mother, but the Clinical Center would not approve bringing drinking problems into the hospital, and I shifted to schizophrenia. So I became known clinically and in the literature for the work with schizophrenia while I continued the interest in alcoholism on a small-scale private practice level.
>
> *Bowen, n.d.a*

Bowen had again encountered societal obstacles in trying to treat alcoholics, this time at NIMH. Alcoholism was still seen as a moral deficit.

> One thing stood out with the work on alcoholics. As long as attention was directed to schizophrenia, the various approaches stood in highest social favor. No method was too radical or too daring. There was only social praise for interest in these problems. Very soon after starting with alcoholics, the social condemnation began to pile up. No matter how successful the treatment effort, the attacks mounted.
>
> *Bowen, n.d.a*

Bowen's ability to think outside the normal confines of research showed at a brainstorming session in August 1954 to stimulate ideas among the researchers who studied schizophrenia. His friend and colleague, Louis Cholden, asked Bowen for suggestions on a project, as they would share ward space. Bowen replied he had no ideas as the study of schizophrenia was not interesting to him. He went on to say that if Cholden wanted to bring in schizophrenic patients and their mothers, he could be interested (Bowen, 1972a, 2). That is how Bowen's research at NIMH began. This idea was foreshadowed in a mother-daughter pair he had seen at Menninger early in the year, March 1954.

Bowen's interest was in building a theory of human behavior, not in the study of schizophrenia. But the intersection of mothers and schizophrenic daughters intrigued him. A plan emerged. Bowen had identified that facts of a human relationship comprised a more scientific approach than accepted assumptions. There was promise to learn more in a method that put family members together.

This could extend his theory building. The mother–child symbiosis could be studied in an environment that, if set up correctly, would not interfere with the ability to observe that relationship. Bowen says he had "spent years thinking about how to do it" (Bowen, 1972b). While he believed that the alcoholism symbiosis was strongest for studying the phenomena, "schizophrenia was chosen as second best" (Bowen, 1983).

Bowen used a standard scientific method of questions, hypotheses, experiments, observations and analyses but without intent to prove anything. Bowen's project at NIMH started with an idea to study the actual relationship between mother and offspring and to define what would meet scientific standards of practice and effective treatment choices. It was a true search for clues, where family relationships were the module of study and the underlying research premise was evolution. A week after this discussion with Cholden, Bowen planned the project for the ward. With a partition installed, Cholden would have one section and Bowen the other. Because it was a female ward, mothers and daughters with schizophrenia would be the patients. Funding was available for this. In a September 1954 letter to a colleague at Menninger, Bowen is heady with the planning.

> The last three months has been one of the most stressful and profitable of my entire life. That puts it too mildly. It has been the most important period of my life …. Most of my enthusiasm these days is about a new project that will start when we open the next ward in October. One day, I made a casual remark about bringing in schizophrenic patients and their mothers, both on the ward together. Before I knew what was happening, the place had taken me up and the plans began to take shape.
>
> *Bowen, 1954a, 2*

My journey through the archives has yielded surprising finds such as an advertisement for McCall's Magazine (Figure 10.1) that appeared in *Time* magazine (Time, Inc., 1954) in the same month that the research project was being configured and ward formed.

There were no written notes on this page (or on any page in the magazine). The term "family unit" in the ad suggests the term was common in American culture. It would be interesting to know what informed the ad agency's presentation. Bowen tended to save materials he considered important to his odyssey. He had saved the magazine with this ad, perhaps as an item of whimsy for a future researcher. Regardless, it fits with the chronicle of his shift from individual to family theory.

Fall 1954

In September, T. P. Rees visited the United States and met with Bowen and other staff on September 2, 1954 (Bowen, 1954d, 1). As medical superintendent at Warlingham Park Hospital in Croydon, England (1935–1956), Rees was well known for unlocking wards and setting an environment for responsible behavior

FIGURE 10.1 Family as a unit advertisement for McCall's Magazine (Time Inc., 1954)
Source: Republished courtesy of Meredith Corporation.

(Maclay, 1963) by the time Bowen was doing his regression work at Menninger. Warlingham Park Hospital was recognized as the first "mental hospital," being opened in 1904 (earlier institutions invariably used the term "mental asylum"). Rees made this institution known for having an open-door policy (Roberts, 2001). In 1954, it became the first hospital offering psychiatric nursing care to out-patients (Roberts, 2001). Bowen credits T. P. Rees for support of his approach at Menninger. "The non-mothering approach, or said the other way, the effort to give responsibility to the patient, is like the approach of Rees and others, especially in England, who have been able to operate mental hospitals without locked doors" (Bowen, 1957b).

It is reasonable to assume that Bowen's project was a topic of conversation between him and Rees, who left a profound impression on Bowen. "At this point I call him the most mature person I have met ..." (Bowen, 1957b, 1). It was enough of an impression that Bowen considered a two-month tour to visit Rees in England (1). By 1956, Bowen hoped that Rees might join his staff that November (Bowen, 1956), although this never transpired.

Another event is of interest because it stands in contrast to Bowen's soon to open NIMH project. The Washington Psychoanalytic Institute approved Bowen's request for trainee status on October 30, 1954 (Halperin, 1954). Bowen had extensive training in psychoanalysis and records of his control cases exist. While he was gathering data for a more factual theory, he continued to practice in the then-operative psychoanalytic theory. This was an area for learning as well.

Chapter 11 will discuss the opening of the project, the families and the project's hypothesis and ward design. Bowen was the chief researcher on two projects. The smaller study, of alcoholics, continued to use the regression approach, modified with the on-site involvement of family members.

From November 1954 to December 1955, the larger family study observed two family members in two separate families in the research setting. A third mother-daughter pair joined the project from January 1955 to October 1955. Bowen's years of reasoning about family culminated in an opportunity to observe direct family interaction 24 hours a day. It was a holistic approach incorporating research subjects, milieu and a broader environment oriented to innovation. "In complexity science this is called emergence, a term used to describe things that are unpredictable, which seem to result from the interactions between elements ..."

Note

1 Thomas S. Kuhn was a physicist and American historian of science who described the elements of shifts in scientific paradigms as related to anomaly induced crisis leading to a new paradigm that better explains external reality.

References*

*Unless otherwise specified, the works of Murray Bowen included in this chapter are from the Murray Bowen Papers. 1951–2004. Located in: Modern Manuscripts Collection, History of Medicine Division, National Library of Medicine, Bethesda, MD. The Accession, Box and Folder information differs for each one and that is included here.

Bowen, M. (n.d.a). [Letter]. (Acc. 2007-012, Box 2, Folder Misc. correspondence).
Bowen, M. (1953). [Annual report of the Bowen Section of D.A.P. for the Fiscal Year July 1, 1952 to June 30, 1953]. Copy in possession of the Bowen family.
Bowen, M. (1954a). [Letter to Prescott Thompson, a colleague and later director of the Menninger out-patient clinic, July 16, 1954]. (Acc. 2007-012, Box 4, Folder Thompson, Pete).

Bowen, M. (1954b). [Letter to a Menninger patient on July 10, 1954]. (Acc. 2007-073, Box 3, Folder Correspondence/Patient July 10, 1954).

Bowen, M. (1954c). [Draft of letter to Dr. Karl, September, 1954]. (Acc. 2007-012, Box 3, Folder Menninger correspondence).

Bowen, M. (1954d). [Letter to Prescott Thompson, a colleague and later director of the Menninger out-patient clinic, on September 8]. (Acc. 2007-012, Box 4, Folder Thompson, Pete).

Bowen, M. (1955). [Clinical record summary, July 1954 to Feb. 23, 1955]. (Acc. 2006-003, Box 3, name redacted).

Bowen, M. (1956). [Letter to Leslie Osborn on July 27, 1956]. (Acc. 2006-003, Box 7, Folder Osborn, Leslie A., Wisc. Psychiatric Inst.).

Bowen, M. (1957a). Treatment of family groups with a schizophrenic member. Chapter 1 in *Family therapy in clinical practice*. New York, NY: Jason Aronson, 1978.

Bowen, M. (1957b). [Letter responding to a request for information on using regression therapy on December 2, 1957]. (Acc. 2006-003, Box 6, Folder NIMH correspondence Q through Z).

Bowen, M. (1959a). [Intrafamily dynamics in emotional illness]. (Acc. 2004-013, Box 4, Folder Presentation 3/31/59 to annual meeting of the Amer. Ortho. Assoc.).

Bowen, M. (1959b). [Summary of psychotherapy with Mrs. (name redacted) July 1954 to September 1955 on August 31, 1959]. (Acc. 2006-003, Box 3, (Name redacted)).

Bowen, M. (1960). [Response on April 7, 1960 to another psychiatrist treating the husband of a person previously seen by Bowen]. (Acc. 2006-003, Box 3, Folder (clinical name redacted)).

Bowen, M. (c. 1965). [Draft. Family psychotherapy with schizophrenia in the hospital and in private practice]. (Acc. 2007-073, Box 3, Folder File name: Working papers (book chapter on family psychotherapy and schizophrenia).

Bowen, M. (1972a). [APA interview with Mark Johnson on February 2, 1972]. (Acc. 2007-012, Box 2, Folder APA interview: Dr. Bowen and Dr. Johnson).

Bowen, M. (1972b). [Notes from interview with Merrill Secrest on February 1, 1972]. (Acc. 2007-073, Box 1, Folder Merrill Secrest).

Bowen, M. (1983). [Draft of presentation on family systems and alcoholism, June 9, 1983]. (Acc. 2003-044, Box 5, Folder VA beach, VA).

Bowen, M. (1985). [Draft notes for presentation]. (Acc. 2003-044, Box 6, Folder Ortho – NYC April 24, 1985).

Cohen, R., & Jenkins, W. (1954). [Administration and Research Planning of Clinical Investigations Program, analysis of NIH Program Activities, Project Description Sheet]. (Acc. 2006-003, Box 4, Folder NIMH–Annual report (project description)).

Halperin, A. (1954). [Letter from Alexander Halperin to Murray Bowen on October 30, 1954]. (Acc. 2003-044, Box 1, Folder Name redacted).

Kuhn, T. (1970). *The structure of scientific revolutions* (2nd ed.). Chicago: The University of Chicago Press.

Maclay, A. (1963). Address at the memorial service to Dr. T. P. Rees, August, 1963. *Mental Health, 22*(3), 118–128. Retrieved February 8, 2021, from www.ncbi.nlm.nih.gov/pmc/articles/PMC5083428/?page=1.

Roberts, A. (2001). Index of English and Welsh lunatic asylums and mental hospitals. Asylums Index 2001. Retrieved May 17, 2020, from http://studymore.org.uk/4_13_TA.HTM#T.P.Rees.

Time Inc. (1954). [McCall's advertisement in Time magazine, LXIV, (7), August 16, 1954, p. 31]. (Acc. 2006-033, box 2, Folder "TIME" magazine Aug. 16, 1954).

Westley, F., Zimmerman, B., & Patton, M. Q. (2006). *Getting to maybe*. Toronto: Random House of Canada.

11

THE BACKSTORY ON THE RESEARCH PROJECT

November to December 1954

Life at the Clinical Center

The Center's design supported Bowen's intention to create a close-to-normal environment. Though it was a clinical research center—basically a workshop for human studies—administrators carefully planned the general milieu to give a sense of normalcy for patients, families and visitors. It offered community perks to normalize the lives of the study subjects. There was a cafeteria for meals together, hair salons for men and women, a newsstand and a small retail store selling general merchandise (Weinstein, 2013, 126).

Patients and family members at the Center could move between floors unless they were on a locked unit (in which case staff would accompany the patient on any trip off the unit). There was a general recreation floor, where staff and patients held musical revues, entertainment and parties in the auditorium. On this same floor was a nondenominational chapel, a solarium and two sun decks (Weinstein, 2013, 126). Occupational therapy (OT) and recreational therapy (RT) were standard practices on psychiatric units, and social events sponsored by the hospital were available.

Bowen and Cholden's patients were on 3-East (3-E), one of the Center's 24 nursing units. One nurse's station served both Cholden's and Bowen's projects. There was a large gathering area for daytime activities and some ward meetings, with space for a dining table so family members could have meals together. Two nurses or aides shared meals with the families. The dietary department supplied food, but family members could request food items if they wanted to cook. It was common that certain family members would make desserts or cakes for special occasions. Anyone using the kitchen also cleaned it (Kvarnes, 1957).

In Bowen's section, there were seven double-bed bedrooms for family members (Bowen, 1958). They served as additional living space as each room had a desk, armchair and ottomans besides the bed (Kvarnes, 1957). The housekeeping staff

DOI: 10.4324/9781003027287-12

did basic cleaning with families responsible for their own belongings and any mess they created.

The ward environment at NIMH

Bowen adapted his research to the context of a split ward. Cholden's project also studied women. Females and schizophrenia fit better with the ward space than would mixed genders or a parent and offspring with alcoholism. Other doctors could admit patients to the unfilled beds. Sometimes this was disruptive, and it certainly tested the resilience of the nursing staff to implement various physician's expectations. Cholden had his own instructions for the nurses regarding his project. Another doctor using a bed for their patient had his or her own protocols for the ward nurses to follow. Nurses implemented Bowen's specific instruction 24 hours a day.

In Bowen's section, a particular symbiotic relationship was the object of study. The symbiosis was to be overt and positive (Bowen, Dysinger, Brodey, & Basamania, 1957). Positive meant the mother and daughter tolerated each other and overt meant their interactions with each other were observable. Researchers' attention went to the subjects functioning in the environment, the symbiosis as it operated between the two and not on the manifestations of schizophrenia in psychotic symptoms.

Once the project was in operation, the nursing staff left caretaking between the two family members. The nurses were assistants to family members, observers and recorders of family interaction. Bowen trained the nurses to do this as it went against a nurse's nature and challenged their professional standards. They had to buy in to the plan.

Lyman Wynne, a psychiatrist, psychologist and psychoanalyst, whose office was across the hall from Bowen's, is quoted in a 1990 memorial article on Bowen on the atmosphere at National Institute of Mental Health (NIMH) in the mid-1950s.

> Murray was only one of a number of mavericks there. Almost everybody was on the outs with the orthodox psychoanalytic establishment of the time, and anybody interested in the family who paid much attention to analysis was considered a kind of enemy. It was all very loose and lively, and you could do whatever you wanted as long as it was interesting.
>
> *Wylie, 1991*

Bowen's initial research questions

Bowen's two concurrent projects, families with either alcoholism or schizophrenia, had one goal: <u>understanding</u> severe symptoms as being connected to relationship processes. Both projects looked at familial connections; and the adult, either spouse or parent, was the therapeutic agent of change, not the doctor or staff. This next iteration of Bowen's research studied in vivo a family twosome, either horizontally with husband and wife, or cross generationally with parent and child and two different symptoms. Bowen's in-patient NIMH project on alcoholism extended the regression approach with the treatment modification of involving a spouse, not nurses as caretaker. It began in August (Bowen, 1955a) and was in place as the family project with schizophrenia launched in November.

There were two tracks from which Bowen operated: psychoanalytic theory and a, as yet not named, natural systems theory. Though orienting more and more to the latter, he would not discard any body of knowledge. He expected that his work with psychotic-level problems could extend psychoanalytic theory toward science. His expansive interpretation of symbiosis considered either positive or negative over-connection worthy of study. He chose to start with positive over-attachment, as the negative variety could not tolerate living together. The nature of the relationship's intensity suited Bowen's psychotherapy approach, to see if the symbiosis could resolve itself in a specific environment and studied in a hospital setting (Simon, 1980).

Project staff for the schizophrenia project

Bowen was the principal investigator. It is unknown whether he chose or the administration assigned the other psychiatrists, primarily Dysinger. Mrs. Thais Fisher, a social worker, was placed there by the Social Work Department. The Nursing Department assigned the nurses. The nurses and attendants worked in eight-hour shifts, seven days a week. "The NIMH project's three-shift staff consisted of a mixed-gender, mixed-race nursing team of 13, including two staff nurses, three team leaders, a head nurse, nursing assistants and a unit clerk" (Kvarnes, 1959). Four other staff who did not rotate shifts were an OT and the three active clinical investigators (Bowen, Dysinger and Fisher). A Research Assistant (silent observer) was also on the ward.

Robert Dysinger, M.D.

Dysinger's own explorations started before joining Bowen's project. He studied relationship changes accompanied by emotional disturbance and internal physical symptoms and wrote an unpublished paper "Acquired Hemolytic Anemia Associated with Acute Schizophrenic Psychosis a Clinical Note" (Dysinger, 1956). The draft summarizes the clinical data from one year of psychotherapy with a 19-year-old woman, seriously ill with blood problems, who became psychotic. Its conclusions suggested a link between the onset of physical symptoms and major shifts in important personal relationships, postulating that anemia was an alternative response to psychosis. The potential strength of the psychosis influenced the process of the anemia. His early involvement in Bowen's project was as a distant yet interested observer, willing to offer comments if asked (Dysinger, 1989, 3). In midsummer 1955, when the dawning recognition of family as an emotional unit shaped the research hypothesis toward a new clinical focus and treatment plan,

Dysinger became an engaged participant (3). His major contributions to Bowen's investigations were his work on the influence of symbiosis in doctor/patient relationships (in 1956 he set up a medical clinic within the ward); on the relation between an emergence of symptoms and lack of a replacement for the other when a relationship ruptured; and on tracking the broader context of anxiety transfer between the NIMH system and the operations in Bowen's unit. Dysinger was the only investigator aside from Bowen who was on the project from its launch until its termination. Significant correspondence exists between Bowen and Dysinger through 1989. He died in 1993.

Mrs. Thais Fisher, ACSW

Mrs. Fisher worked on the research project as its social worker from November 1954 until October 1955, when she left on maternity leave with no intent to return. Her casework reported on meetings with the mothers and gave detailed information on their conversations. Those notes showed how she applied the dictates of Bowen's ward policy, her reflections on her own responses, and her struggles to be present for very challenging mothers. These reports bring alive the reality of establishing a relationship "to the mature side," which often took hours of engagement to emerge.

The project's hypothesis

In its simplest form, the hypothesis stated "Manifest symptomatology, in the patient, psychosis; in the mother, psychosis and other symptomatic modes of adaption, appears when the gratifications of their needs are threatened" (Bowen, 1956a). "If the environment can treat the patient as an adult, there is a chance for recovery" (Bowen, c. 1957). Underlying the hypothesis was an interactional view of relationship adaptation to the environment, along with the idea that family's underutilized strength could improve treatment. "The short-term hypothesis for each sub study in the total research" related to what made up a healthy environment in which family members could learn how to address family problems. The hypothesis was

> based on the theoretical assumption that the mother and patient could resolve the emotional attachment if both were together in a specific supportive milieu in which NO ONE WOULD TAKE SIDES EMOTIONALLY OR TAKE ACTION FOR EITHER AGAINST THE OTHER IN THE INTENSE EMOTIONAL FIELD BETWEEN THE TWO [Caps in original]. It was hypothesized that these two had lived with each other for years, that each knew the other well, that neither had built-in mechanisms to automatically control the other and that neither had ever seriously hurt the other [underline in original] .... This one principle remained an unaltered core principle of the research and it is still a core concept in the family theory developed after the initial research. The concept that a tension system between two people will resolve itself in the presence of a third person who can avoid

emotional participation with either while still relating actively to both is so accurate that it can be predictably repeated in family psychotherapy with less severe emotional problems.

Bowen, c. 1976, 1–2

And there was "… a broad, long-term hypothesis to govern the total study over the years," referencing the inductive leap to the family as a naturally formed unit (Kerr & Bowen, 1988, 329). Emotional illness reflected a disturbance in the emotional system, an inner guidance system shared by lower forms of life in the evolutionary past of living things. Bowen inferred that natural laws govern this development (Kerr & Bowen, 1988, 329). While this was in place from the Menninger work, he extended it in 1955 with the crucial evidence of the family as a unit.

Based on the scientific principle that it is profitable "to first study the gross aspects of a poorly understood phenomenon before proceeding to microscopic detail" (Bowen, 1957, 3), the ward was Bowen's laboratory of human subjects. A research setting that allowed family members to be hospitalized offered potential for recording new observations, and nursing staff played a central part.

A central principle was that staff keep its hands off the symbiosis, to trust the thesis that the symbiosis knew the direction to go to resolve its dilemma …. The staff was there to aid in a growth process and not to direct it.

Bowen, n.d.a, 1

The belief that each individual's growth tendencies surface if support is available is clear in the early papers from this period. Making support available to families, not taking over for the families, fostered maturity. With support of the mature side in each individual, the mother would lead the change for the daughter. Natural inclinations, a core belief that all living things grow, surfaced in these conditions.

The criteria for admitting families to the ward

The Center charged no fee for families to be in the program; it was a government-funded project.

The strategy was to bring mothers and their daughters with a previous diagnosis of schizophrenia to the research ward. The intensity of the mother/child contact, not the magnitude of the schizophrenic symptoms, was the criterion for admission. Bowen studied families, not a diagnosis. Mother and daughter's thinking, feeling and relating time focused on the other. They could tolerate being in the same space. The mother was to spend as much time as possible with the daughter, even living with her in the same room if she wanted, in keeping with the project's sanctioning of closeness for its value in making people comfortable.

During the preadmission interview, Bowen and the social worker met the mothers and outlined the minimum and maximum that the project would offer to families. They received no promises or expectations about outcomes beyond that some people may benefit, some may not. What they would get is the time and attention of researchers and staff interested in what might make a difference in the relationship. The staff would do what they could to understand the families' problems and to allow access to the staff's observations and records for the family's own educational use. Families could attend and take part in staff meetings, though early on none attended.

The contract stated that help was available, and the family committed to work at the problem without time limits, while the program proved beneficial. Continuation in the project was a decision for mother and daughter. Participation was possible as long as both found it helpful.

Compared to a negative symbiosis, where the two family members avoid being near each other, a positive symbiosis describes an interfering mother who shows positive over-involvement. The mother's presence at the hospital demonstrated her accepting that she was part of the problem in the relationship. For research purposes, the symbiotic intensity was most important. Bowen built in a liberal leave policy for a family member, recognizing the potential strain of daily living contact. He expected that mother and even daughter might want time away while on the project.

> Even a healthy family has emotional crises when people are too long in closed living space. The crises are more frequent in impaired families. The leave policy permitted some to get away for a time until feelings subsided. It was implemented for the families, not for research needs.
>
> *Bowen, 1986*

Implementing the program

Training and exercises with nurses and aides prepared them for the very different experience of being present on a ward as a source of aid to be drawn upon in contrast to their typical caretaking capacity. Bowen intended the exercises to reduce "observational blindness" in staff (Bowen, 1965, 105).

Staff training emphasized a relationship with the adult side of mother and daughter while respecting the infantile striving for caretaking. The nurses were to avoid motherly responses, as the actual mother was there. They could offer time, an honest effort, a presence and a respectful attitude.

All staff, including the psychiatrists, social worker, social science analyst, nursing and rehabilitation professionals, were to develop an orientation that emotional illness showed a particular relationship process. The symbiotic relationship was not a pathology or an individual deficit. The crucial attitude was that the intense attachment inhibited the ordinary path of development; it was not a permanent condition. The attachment could resolve and the patient moved to mature growth under the necessary conditions that the project provided. An approach based on the attitude that the attachment was a pathology would establish a different (and less effective) emotional atmosphere.

Nurses dealt with immediate and daily problems that might arise. Staff were active in paying attention to boundaries between self and family (Bowen, 1995, 37).

> Nurses were assigned to the unit by the nursing department at NIMH. Over the course of 24 hours, they observed the families whenever they were in common areas of the ward, when invited to accompany family members to activities off the ward, or whenever nurses were in physical proximity to family members such as in the family members' private rooms. There was an intercom in the private rooms that connected to the nurse's station that was under the families' discretion to leave on or turn off. The nurses ate meals and participated in ward activities with the family representatives. Families were informed of this level of observation prior to admission.
>
> *Rakow, 2016, 142–143*

Staff were to stay in the present, seeing the back and forth of the mother–daughter behavior. Anticipating the strong pull on nurses and aides to get involved in the mother–daughter interactions, the training featured ways of managing self, observing the characteristics of these cycles and remaining neutral. Other operating principles respected the capacities and responsibilities of family members and supported their accepting them. "The staff define specifically the boundaries of the structure and the family members determine their course with the structure" (Bowen, 1995, 37).

Staff kept records of what they saw and heard. They were to use descriptive words in their writing and speaking, not psychiatric jargon. The intention was to help the staff understand how the words one uses align one's self with another—ego merger. They were to monitor self to recognize a process of identification with family members. This occurs when support goes to one member while withdrawing from another. It took time for the staff to learn to not interpret the behavior. The nursing chart shows their increasing skill as they implemented this approach. Bowen held daily staff meetings to help with project challenges. By generating an environment based on the idea of responsibility, Bowen expected staff who felt challenged to rely on self, not on hospital structure, for their position. For example: A response from within self would start with "I," not "the doctor said, or the hospital requires." Staff were to help without giving advice or directing what a family member does (Bowen, n.d.b). And most importantly, staff were to leave the symbiosis where it originated. By keeping the transference within the family, staff could observe it and learn how it operates.

The operating principle was that the family's use of its resources, not the staff's interventions, brought about the change. The staff's contribution was to create a growth-stimulating situation. Bowen intended the round-the-clock observations of mother and child to allow for further refinements in managing the environment in order to increase maturation opportunities. The ward would not be a static setting. An enriched environment of professional staff, nurses, aides and activity therapists was available to family members for help upon request.

Bowen taught the staff the conditions to watch for that he had identified in the Menninger investigations (forced mothering, environmental demands, demands from within, unpredictable environment) and how they triggered symptoms. Bowen's project depended on the design of the Clinical Center, the setting of the milieu and on the staff's capability to juggle dual loyalties to the administration and to Bowen's protocols.

Extending and modifying the non-mothering approach

The mother would be the agent of change for the daughter. Consistent clarification by staff would be important when the mother did not consider herself part of the problem. A new awareness of self would be the first step to behavior changes in self. Clarity was possible through the staff relating to her but not through staff enforcement. Mother could use her awareness of self to change. Reciprocally, this would bring changes in the other, or a push to change back.

While staff observed families for investigative purposes, it also opened communication between staff and family members. The nurses could comment neutrally on what they saw going on in front of them. For example, observing an intense exchange, a nurse could say, "I can see how each of you has a hard time being heard by the other. It's a real problem." If pressed to solve the problem, the staff had options other than giving advice. Further listening until an adult response emerged from the family member was one option. It took patience. The point was to maintain a relationship without moving toward or away from the person.

Bowen knew that familial contact between people over many years meant there was a pattern of habituated or repeated responses between household members. This was one element of schizophrenia. He reasoned that a more favorable environment coupled with new experiences could interrupt these repetitive responses, forcing fresh responses and challenging a family member to think or act differently. The project opened with a mother and hospitalized child in an environment that maximized personal responsibility for each of them and of the staff. Each family member had a significant other to turn to in Bowen's project. The mother had a social worker, and the daughter had Bowen as a therapist, connections that modeled independence. Each staff member offered an element of diversity to the environment as they developed their own ways to relate without attaching. Staff was not to usurp a family member's personal responsibility.

Milieu

The milieu integrated two key principles: the staff were resources, not caretakers, and activating a family's own inner strengths was important. Either mother or daughter, or both, could choose, or not, to engage the resources offered in the milieu. The design of the milieu was to substitute the actual mother as caregiver who meets daughter's needs with supportive, neutral nurses and a resource-rich environment. The detachment of the staff modeled for mother and daughter how to recognize and incorporate self. Reflecting on the program, a line in Kipling's poem "If" comes to my mind: "If all men count with you, but none too much." In fact, I later found the poem among Bowen's papers (Army materials, 1945).

The ward structure served two purposes. It created a climate of people learning from each other and acted as a precedent for observational learning from family to family.[1] Researchers and the families both could observe the interactions. The milieu was a reversal of the usual hierarchy in a hospital, where staff are the experts. The way a nurse was trained to spend time with mother gave her a measure of intimacy and familiarity and modeled how to relate with a daughter without reactivity driving the relations. Through learning this, mother transferred it to helping her daughter. It was not an easy sell to staff. "At first other staff members (were) reluctant to go along with the idea of family 'living in' with patients" (Bowen, 1956b).

Daughters were hospitalized. Mother exercised a live-in option. The therapeutic environment came from staff's curious, nonaligned attitude and noninterference in the intense parent/offspring relations, and from being present and available to help when the mother or daughter asked for it. This put additional responsibility on the staff. If nurses or aides perceived a request as wanting them to take over for the family—the phrase "emotional bid" was used to describe this action (Bowen, 1959, 1)—they could say no. The family member wanted someone else to figure out a situation and clue her in on what was going on in the family. By not doing this, the staff demonstrated that the individual herself could be more objective "and see the forest instead of the trees" (Bowen, 1959, 1).

When declining requests they considered inappropriate, staff had to quickly assess the reality of the petition. Were the family requests coming from a position of dependency or from actual need? Staff assessed their own responses as well. Did they have an attitude that the person was helpless or that the person had inner strengths? Adjusting one's own attitude created a more differentiated environment. Assessment and observation of self and the family members was constant, illustrating the principle of not expecting something of another that one has not already applied to self.

The loose structure of Bowen's unit was intended to simulate a natural setting in an institution. It was also to assist the researchers in watching how the mothers and daughters managed their relationship in a nonparental or non-authoritative environment. Parental authority rested with the mother, the actual parent, not with the staff. In this way, family issues could surface and be addressed in the open. Bowen's energy went to training the staff on these intense relationships based on what he knew from Menninger. He predicted problems previously seen, such as the closeness/distance cycles in the Menninger regression explorations. To avoid surprises, Bowen did exercises with staff on how such a cycle displayed and talked through their possible responses. This built in security for staff and stability for the project. Mothers assumed responsibility for daughters' behaviors, in compliance with hospital rules. Mother and daughter worked out their own relationship. Advice and direction were off limits for staff, but help from staff in thinking through problems was allowed. Mothers directed the treatment. Staff made decisions around treatment in consultation with mothers and patients. From the work at Menninger's, Bowen knew that an unpredictable environment could trigger psychotic symptoms in an impaired person. The effort on the ward was to have a predictable environment that offered a variety of supports from doctor, nurses, nurses' aides, social workers, OT and RT.

Treatment plan

The treatment offered was supportive therapy for the parent and individual psychotherapy for the impaired offspring. It was a hybrid of the earlier regression and non-mothering practices. Mother could have as much time with the social worker as she chose. The relationship was a way for the mother to address her own needs, giving her a person to draw from or a relationship to meet her emotional needs other than from the child. Knowing that a symbiosis is transferable via mothering, supportive therapy emphasized continuing the relationship rather than solving the mother's problems. Personal matters were not the focus.

The daughter set her own appointments with the psychiatrist. She could work out her inner concerns in meetings with Bowen as often as she wanted. In these sessions, Bowen recognized the infantile side of the daughter without relating to it. When the daughter asked, "Am I hopeless?" Bowen replied, "Some do recover, some don't. I don't know how it will go with you" (Bowen, 1956c). The therapist was not the parent. It was possible to do this when the parent had daily contact with an adult child. This is the family version of the work done at Menninger's, only here the patients and stand-in parents were the real members of the family.

The staff were not to interfere in the interactions between family members, but they were to relate to the mature part of each. This kept the interdependency between the two family members. Leaving the intense interface between the two was intended to keep the therapeutic relationship with psychiatrist and social worker on a higher level.

This treatment approach supported the central idea in the hypothesis that the mother is the therapeutic change agent for the daughter (Bowen, 1956b). The daughter gains in her functional level when the parent does not depend on her, the basis of working toward maturity. Responsible mothering assists maturation. When the mother can first recognize this in herself, then ease the inner need for her daughter, the growth instinct can activate in the child. Improvement in family relationships came from their efforts. Said another way, progress belonged to the family. It was not the staff's responsibility. The staff created the climate and provided the support to nurture this. The family made use of the situation and the resources for their own progress. From the first interview, the mother stated how to proceed, first identifying where she would live, whether she would live with her daughter or instead live nearby. The principle was "… the belief that parents would add something to the treatment … in essence we wanted her to say how she thought she should proceed in this treatment effort and how she wanted us to help her" (Bowen, Fisher, & Bowe, 1955, 5). The treatment offered each family member the same choice of drawing from another relationship, beyond the one with each other.

Psychiatry knew little about how human symbiosis operated beyond normal developmental phases or what emotional variables sustained it. This was an unstudied area with actual family members. The NIMH study observed the biological parent and impaired child in symbiotic attachment, checking for resolution of the symbiosis in vivo, in a richly supportive, neutral environment. Their living together would allow close monitoring of the symptoms of schizophrenia, in addition. Staff would see two people merging or fusing into one, thinking, feeling and being alike, and then, unable to sustain this, rejecting each other. The reduced anxiety about separation of the mother/child pair would be a prompt for the symbiosis to dissolve. With consistency of feeling less anxiety, their confidence would increase and individual growth would spark spontaneously. Change of this degree took time.

> At the start, this particular type of symbiosis was thought of as rigid with fixed personality fusion. It takes time to think about a life course, to decide on changes and to implement them with fortitude. The expected length of the project recognized that it takes space to change one's self within one's relationships. A goal was to continue five to ten years if needed … the mother further detach herself, then the patient again shift. The process repeating until each could grow free of the other, and each in an autonomous mature interpretive psychotherapy relationship.
>
> *Bowen, 1957*

Providing a caring, neutral someone for each mother and daughter could ease the eventual separation. A cycle of oneness to twoness is universal in living things. "The oneness-twoness cycle is much more definite and more easily completed in lower forms of life than in man" (Bowen, n.d.c). Others, interested in science and application of the theory, have extended this from "the cycle is oneness-twoness to a more complex oneness-twoness" in humans because of the human brain (A. Post, personal communication, August 23, 2020).

Record keeping and other administrative considerations

A strong data trail exists in this project. Staff kept three shifts of multiple records: admission write-ups, intake assessment summaries, laboratory reports, discharge notes and the social worker's process notes of meetings with the mother, all conforming with hospital policy. The Research Assistant created daily and weekly charts of pertinent information culled from those records. She did this to aid accuracy and avoid distortions of what staff had recorded as seen and heard. These charts served as the basis for the theoretical discussions among staff. The social worker kept her session summaries for administrative record keeping compliance and for research examination. If a mother was activating her inner strengths, it would be noted there. Bowen recorded the clinical interviews between him and the daughter in the first year. Several audiograph recordings of these interviews remain among the archival materials. "Bowen accepted that all writers, regardless of professional status, could equally gather the data if they understood the concept. These notes make up an exceptionally detailed record of the project" (Rakow, 2016, 143).

Requirements of the larger hospital had to be met along with those set for the project. Hospital policy designated the mother to be a normal control. Anytime a parent in the project left the hospital, whether the parent lived on the unit or in proximity, spending daytime hours at the hospital, the departure was treated as a discharge, with its attendant paperwork. The parent's return to the project was treated as an admission.

Exploring Bowen's archival materials, one can find the quantum leap to his theory of the family as a unit in his observations, actions and clinical practice before he put it into words. The freedom to operate in a research-oriented organization made the difference. "Confronting the same constellation of objects as before and knowing that he does so, he … finds them transformed through and through …" (Kuhn, 1970, 122).

Conducting this study was part of Bowen's differentiation put into action. Bowen had a foothold in individual theory by having individual relationships between therapist and patient and between social worker and mother. And he was moving closer to systems thinking by having 24-hour observations, creating a richly sourced milieu and allowing growth to germinate while leaving the unresolved attachment where it began, between the parent and child.

Bowen's far-reaching view of schizophrenia was, it is "… not a clinical entity unto itself, but rather one of the more severe symptom complexes in a long continuum, and one that encompasses almost every experience in human living …" (Bowen, 1955b). He reasoned that the psychosis developed when the offspring was "AWOL" from the relationship in an attempt to grow up (Bowen, n.d.a). It could help to explain the incidence of psychosis in young adults going off to college or trying to live on his/her own.

From the start, the project used the concept founded in his previous observations that emotional intensity between two people dissolved when a neutral third was present. Though establishing that neutral third entity via the staff was only minimally successful in the first months of the project, the effort led to a remarkable catalyst for new advancements, as we will see in Chapter 12.

Note

1 The concept of observational learning is attributed to the behaviorist Albert Bandura and his 1977 studies. https://carpresourceguide.weebly.com/observational-learning.html. Bowen integrated this idea in his research project in 1954.

References*

*Unless otherwise specified, the works of Murray Bowen included in this chapter are from the Murray Bowen Papers, 1951-2004. In: Modern Manuscripts Collection, History of Medicine Division, National Library of Medicine, Bethesda, MD. The Accession, Box and Folder information differs for each one and that is included here.

Army materials. (1945). [Summary, headquarters 220th general hospital 22 September 1945]. L. Murray Bowen papers, National Library of Medicine, History of Medicine Division. (Acc. 2007-073, Box 4, Folder Army), Bethesda, MD.

Bowen, M. (n.d.a). [Family study clinical research project]. (Acc. 2007-073, Box 3, Folder Working papers from NIMH project re: Schizophrenia).

Bowen, M. (n.d.b). [Draft]. (Acc. 2007-073, Box 3, Folder Working papers).

Bowen, M. (n.d.c). [Handwritten draft]. (Acc. 2007-073, Box 3, Folder Bowen-NIMH-Working papers incl. proposal for project).

Bowen, M. (1955a). [Clinical record, doctor's progress notes, August 22, 1955]. (Acc. 2006-003, Box 2, Folder Name redacted).

Bowen, M. (1955b). [NIMH form, report on research activity of laboratory of adult psychiatric investigations, March 15, 1955]. (Acc. 2006-003, Box 4, Folder Report-NIMH general staff).

Bowen, M. (1956a). [Paper. Formulation of the 3-E family study project July 16, 1956]. (Acc. 2006-003, Box 4, Folder 3-E project).

Bowen, M. (1956b). [Letter on September 14, 1956. Response to an inquiry for a referral to help a young boy experiencing mental difficulties]. (Acc. 2006-003, Box 6, Folder NIMH correspondence H through P).

Bowen, M. (1956c). [Draft. The family and schizophrenia]. (Acc. 2007-073, Box 3, Folder Bowen-NIMH-working papers incl. proposal for project).

Bowen, M. (c. 1957). [Draft. A psychological hypothesis of schizophrenia]. (Acc. 2006-003, Box 8, Folder A working psychological hypothesis of schizophrenia).

Bowen, M. (1957). [Draft. Psychotherapeutic treatment of the family as a unit]. (Acc. 2006-003, Box 2, Folder drafts).

Bowen, M. (1958). [Letter to mother of referral family on January 28, 1958]. (Acc. 2006-003, Box 6, Folder NIMH correspondence A through G).

Bowen, M. (1959). [Letter to a research family member on February 3, 1959]. (Acc. 2006-003, Box 3, Folder Name redacted).

Bowen, M. (1965). Intrafamily dynamics in emotional illness. Chapter 7 in *Family therapy in clinical practice*. New York, NY: Jason Aronson, 1978.

Bowen, M. (c. 1976). [Letter]. (Acc. 2005-055, Box 3, Folder MB position statements).

Bowen, M. (1986). [Draft]. (Acc. 2006-003, Box 8, Folder drafts of "odyssey-in theoretical principle").

Bowen, M. (1995). A psychological formulation of schizophrenia. *Family Systems Journal, 2*(1), 17–47.

Bowen, M., Fisher, T., & Bowe, M. (1955). [Presentation to staff on August 19, 1995]. (Acc. 2006-003, Box 3, Folder name redacted Daughter A.).

Bowen, M., Dysinger, R., Brodey, W., & Basamania, B. (1957). [Family diagnosis problem]. (Acc. 2006-003, Box 4, Folder Study and treatment of five hospitalized family groups each with a psychotic member).

Dysinger, R. (1956). [Draft: Acquired hemolytic anemia associated with acute schizophrenic psychosis a clinical note]. L. Murray Bowen papers, National Library of Medicine, History of Medicine Division. (Acc. 2006-003, Box 4, Folder Robert H. Dysinger, July 3, 1956), Bethesda, MD.

Dysinger, R. (1989). [Letter to a PhD candidate on September 11, 1989]. L. Murray Bowen papers, National Library of Medicine, History of Medicine Division. (Acc. 2007-012, Box 4, Folder Bob Dysinger), Bethesda, MD.

Kerr, M., & Bowen, M. (1988). *Family evaluation: An approach based on Bowen theory.* New York: W. W. Norton.

Kuhn, T. (1970). *The structure of scientific revolutions* (2nd ed.). Chicago: University of Chicago Press.

Kvarnes, M. (1957). ["Orienting nursing personnel to a research project in psychiatry"]. (Acc. 2006-003, Box 4, Folder 3-E project).

Kvarnes, M. (1959). The patient is the family. *Nursing Outlook, 7*(3). (Acc. 2006-003, Box 1, Folder Reprints from within NIH).

Rakow, C. (2016). Learning from the nurses' notes for Bowen's 1954–1959 NIMH project: A window into the development of theory. *Family Systems Journal, 11*(2), 131–164.

Simon, R. (1980). A network genealogy: A history of family therapy in the Baltimore-Washington Corridor, Part III—The protoplasmic world: Conclusion of an interview with Murray Bowen. *Family Therapy Practice Network Newsletter, 4*(1), 1–10.

Weinstein, D. (2013). *The pathological family: Postwar America and the rise of family therapy.* Ithaca, NY: Cornell University Press.

Wylie, M. S. (1991, March/April). Family Therapy's Neglected Prophet. *Psychotherapy Networker.* Retrieved May 24, 2020, from: www.psychotherapynetworker.org/magazine/article/947/family-therapys-neglected-prophet.

12

THE FIRST FAMILIES

November 1954 to February 1955

Bowen came to NIMH understanding that patients did better with family nearby. The clinical testing of this recognition and his other newly formed conceptualizations began in November 1954.

The first three families

His search for research families first took him back to Topeka. On September 16, he wrote to a woman he had seen earlier in the year at Menninger's for a single psychiatric consultation. He told her about his move to Bethesda and invited her (and her mother) to consider being part of his study. "It would ... necessitate your mother coming here to live to be with you. My ward set up will be such that mothers, or fathers too for that matter, can actually live right on the ward where the patients stay ..." (Bowen, 1954a, 2).

This invitation reflects Bowen's assertion that it is the mothering function, not gender that is important in a symbiotic relationship and previews the possibilities of having intact families live together on the ward. The mother responded five days later, declining the offer. She wrote "... since consultation with you in Topeka, (my daughter) has shown a marked improvement" (Bowen, 1954b, 1).

Bowen reached out to a local psychiatric hospital in DC in October 1954 seeking families for his project. Several days later, Mrs. A,[1] whose daughter had been a patient at the local hospital for 14 months, called the Center asking for her daughter to be admitted (Bowen, Fisher, & Bowe, 1955, 4). The mother lived several states away.

The daughter, the younger of two girls, transferred to Bowen's research project on November 1, 1954. The mother arrived on November 3, 1954 (6). Bowen marked the start of family psychotherapy with this admission, writing in 1965,

DOI: 10.4324/9781003027287-13

"My effort to develop family psychotherapy began some eleven years ago" (Bowen, 1965a), when he put the mother and the daughter together for in vivo study.

The second group, the B family, entered the project three weeks later on November 23, 1955. An East Coast welfare agency, from which the family had been receiving services since 1937, had referred them, and they returned to the agency on discharge (Family Services Social Worker, name redacted, 1959). After leaving the project, the daughter and Bowen maintained contact through 1974.

In the December 1954 year-end review, Bowen says his "project (is) to check [the] belief that [the] presence of the mother is beneficial to the treatment of schizophrenia [and] schizophrenia is considered a manifestation of an intense interdependent relationship" (Bowen, 1954c).

The C family, admitted in January 1955, had five children. The oldest sister, age 17, was married and living on her own. A 16-year-old brother was in the army (Bowen, 1962). The third of the five children, a daughter, age 15 and diagnosed with schizophrenia, was first admitted to a psychiatric hospital in early 1954 with confused thinking. She had past difficulties with acting out symptoms. The hospital discharged her for a trial period in September 1954, and quickly readmitted her until transferring her to Bowen's project. Both 14-year-old and 10-year-old sons lived at home. This family was also in the area's welfare system.[2]

The variable of sibling position

Sibling position played a part in the selection process. Mother/daughter pairs had one child, two siblings or multiple siblings. This variance was deliberate, to observe and learn, if possible, how a particular child becomes part of a symbiotic pair (Bowen, 1955a, 1). Family A contained a mother and a youngest daughter; Family B had a mother and an only child; Family C was a mother and middle child with mixed gender siblings. These families, with their multiple representations of siblings, could yield information on the selection of a particular child for meeting needs of a parent.

Family A, November 1954

Daughter A was 31 years old; her mother, 58. Mothers were registered as "normal controls," and she had a choice of visiting daily or of living there on the ward. Mrs. A visited daily. Mr. A, the father, worked for a well-known national organization as a writer and lecturer on his worldwide explorations (Bowen et al., 1955, 11). The As had two daughters born three years apart. Both were born when Mrs. A's mother was ill and dying. The oldest was married and lived in the Midwest. The youngest one resembled her father (4). The couple divorced when Daughter A was 14. Mr. A remarried the same year and had several more children (2). Mr. A visited his daughter regularly during her stay at NIMH (Brodey, 1957). Daughter A became psychotic at age 15. Subsequently, she had three hospitalizations totaling ten and a half years and had received several courses of both insulin and electroshock treatment (Bowen et al., 1955, 1, 3). For a few years, she lived with her therapist on the West Coast. She had been in a local DC psychiatric hospital for more than a year and plans were being made for her transfer to a state institution. "In November 1954, I chose her for a project … to bring mothers into the hospital situation with patients. She was chosen because her mother had no outside responsibilities and could be a part of the project" (Bowen, 1955b). Mrs. Fisher's notes mention that Daughter A "… seemed to have quite a bit of 'spark in her' even though she was psychotic and delusional …" (Brodey, 1957, 5). Again, selection of subjects for the research depended on the characteristics of the symbiotic relationship.

On her first day in the unit, Daughter A found the unstructured environment bewildering. No one told her what to do. She remained waiting in the lounge until midnight when a resident passed through, saying good night as he did so. This cued her to go to bed (Bowen et al., 1955, 1). Daughter A began psychotherapy the day she arrived and continued five-day-a-week, 40-minute sessions until May (6) when she went on leave for the summer (Brodey, 1957, 7). The mother began sessions with Mrs. Fisher on the day of her arrival, too. There were 17 half-hour sessions in September (Fisher, 1954a, 1). Mrs. A met with Mrs. Fisher up to five times a week (Bowen et al., 1955, 7). Neither mother nor daughter used familiar terms when referring to the other. Mother called her daughter "the lady" and the daughter used "My parent" or "she" to refer to her mother (4, 9). Both are examples of creating conceptual space within physical proximity. Daughter A spoke of her father with "affection and pride" (9).

A staff account describes the various stages of Daughter A's first days.

> The first stage is called the admission and orientation stage, which lasted 6 days after admission. In the previous hospital, the patient had been in a chronic rebellion against hospital rules. Here she was bewildered by the lack of structure. She asked hundreds of questions about rules and procedures ….
> When mother arrived, the chronic argument which had been directed at the previous hospital was now directed at mother. The mother was unsure and vacillating. She could not say 'yes' or 'no' in response to her own daughter.
> *Bowen, 1956a, 9*

The second stage for Daughter A, one of confusion and heightened upset, began within a week of admission. This stage had manic overtones of elation, frequent changes of clothes, a babble of rhyming words, numbers and symbols, over affection to staff, excessive cigarette use and one incident of setting a fire in a wastebasket (Bowen et al., 1955, 6). Such incidents induced fear in Mrs. A that the hospital would ask her daughter to leave. Mother acted helpless around her daughter, alternating between anger and indulgence. One of them informed father of the daughter's hospitalization. Within two days of his daughter's admission, Mr. A wrote to the hospital that he intended to visit soon. He expressed willingness to meet for an interview if staff was interested. He visited November 11. During the week of his visit, the nurses captured in their notes that this young woman had a bit of the trickster in her. It was a useful skill to keep the mother off balance (3).

> There were many episodes like the following one: The patient asked mother for her wrist watch. Mother asked about ward rules concerning watches. The patient kept asking. Mother avoided talking to the patient about it. She told a nurse the patient attempted to take the diamond watch off the mother's wrist. The next day, the mother was not wearing her watch. The patient took her mother's handbag, found the watch there and went toward the bathroom. Mother's looks seemed to plead for help. In a few minutes, the patient flushed the toilet. Later the patient returned and after a few minutes of tense silence opened her hand and gave back the watch. The psychosis subsided after ten days.
>
> 6

The arrival of the father's letter, his visit and the timing of the watch incident foreshadows what staff observed about his involvement. Even from a distance, he had an influence on the mother–daughter relationship.

Within weeks of admission, Daughter A began volunteer work, typing for the hospital newsletter (Brodey, 1957, 6).

Relationships beyond the mother–daughter pair

A highly interdependent relationship also existed between the mother and her youngest sister. Mrs. A reported that the sisters' relationship contributed to her sister's divorce because of the sister's husband's jealousy of the relationship (Bowen et al., 1955, 4). The younger sister had a 30-year history of emotional difficulties and periodic hospitalizations. Widowed, she had divorced her second husband, a psychoanalyst who had been her therapist (1).

The tangle of relationships between the two sisters is a useful illustration of togetherness in a family. Each sister had an offer from their ex-husbands to move in with them, to live with the husband's current wife and children from that marriage (2). Mrs. A's brother-in-law, though remarried, even offered to resume treatment with his ex-wife (2). In obtaining the family history, the social worker had to clarify which marriage Mrs. A was speaking of, her own or her sister's. The clinical notes showed that many of the patterns between Mrs. A and her sister repeated in the contact with Mrs. A's child (Fisher, 1954a). This observation captured Bowen's interest, but it was not then possible to hospitalize the extended family. Dr. Bowen wrote

> I wish there had been a way to have made a place on 3-EAST for your aunt because it is part of my goal to work with multiple members of the family in an effort to learn as much as we can about relationships of family members to each other.
>
> *Bowen, 1955c*

In November, Mrs. A's sister placed many demands on Mrs. A, requiring her to prepare their meals in the home and to care for the sister's daughter, Mrs. A's niece. Her sister's helplessness convinced Mrs. A that the sister should not live alone. By the 16th of the month, Daughter A's psychosis had lessened, so her mother took a day off for herself on the 17th. It meant missing her appointment with Mrs. Fisher, not seeing her daughter and not being available to her sister. The sister "ha[d] an excessive nosebleed along with a fainting spell, and it was necessary for her to be hospitalized for a brief time" (Fisher, 1954a). These observations collectively pointed to questions of family process going beyond two people. There was an overlap of functioning within these three members of the family. The marriages of both sisters had multiple similarities in relationship closeness and distance (Bowen et al., 1955, 1). For Mrs. A, relations with her sister had parallels with relations with her daughter. They both demanded her time, and she felt tied to them. Said another way, she could not easily be away from either of them as something of note would occur. There is a theme throughout the notes of Mrs. A wanting others' approval and of feeling anxiety about being judged by her daughter's conduct. If Mrs. A took action for herself, her sister developed physical symptoms. By the third week in November, Daughter A was stabilizing. Mrs. A got physical symptoms following her daughter's improvement.

Staff observed another pattern: When the father visited, Mrs. A absented herself. Mid-month, when the father visited, mother did not visit. Observations of the broader family gave greater insight into the family's emotional life. The visits of other family members with the research families, the impact of those visits on their relationship, along with how the mothers and daughters involved staff routinely in family interactions all pointed to the idea that the dyad was not the primary unit.

Functioning reciprocation

The increase in physical proximity between mother and Daughter A and the development of physical problems in the mother prompted the question, Where was the problem in the family, in whom? Bowen used this family as an example of "functioning reciprocation" (Bowen, 1960, 59) in his 1960 paper, "A Family Concept of Schizophrenia." The paper became Chapter 4 in his book *Family Therapy in Clinical Practice*. He wrote of "functioning reciprocation":

> The 'projection' occurs also on the level of physical illness. This is a mechanism in which *the soma of one person reciprocates with the psyche of another person* [Italics in original] ....A striking series of such reciprocations occurred in a mother, in response to rapid improvement in a regressed patient. Within a few hours after each significant change in the patient, the mother developed a physical illness of several days duration .... These marked reciprocating mechanisms are most common in, but not limited to, the mother-patient relationship.

59–60

The daughter showed positive changes by month's end, taking part in ward and hospital activities and tolerating visits outside the hospital with family. She took initiative to ask her mother for privileges. Mother A invoked hospital rules to avoid answering her daughter's requests. The nurse's notes record that when the daughter did better, the mother developed physical symptoms.

> To the staff, mother said she did not reply to the patient because she was sick and would not understand. She finally worked up enough courage to take the patient for a drive. This was about Thanksgiving of 1954. The patient was a perfect lady. It was their first trip together in more than a year and a half. The following day, the mother 'lost her voice' and stayed at home several days with laryngitis. During the following two months, mother developed a somatic illness for several days following each trip out with the patient. Following the first five or six trips, she developed an illness immediately after, or within a few hours after the trip.
>
> *Bowen et al., 1955, 10–11*

Mrs. A had six different physical symptoms from Thanksgiving through Christmas. These related to increased contact between her and her daughter, whose functioning was improving (6–7). Somatic incidents that aligned with increased physical proximity contributed to understanding the nuclear family's "somatic responsiveness to emotion" (Bowen, 1986). Bowen's alcohol project had the same observation, despite that the patients had a different diagnosis.

Family B, November 1954

Family B[3] arrived two days before Thanksgiving. This mother chose to live in the unit with the daughter. These two were the only pair to live together on the ward from the beginning of the project until discharge three years later, in October 1957. Prior to their acceptance into the project, they lived on welfare and had become wards of the referring family agency. On admission, Mrs. B was 52 years old. The daughter was 17. Mr. and Mrs. B had divorced 15 years earlier.

Mrs. B behaved as if she was more impaired than her daughter. She drew attention to herself. The interactions between her and her daughter ranged from extremely loving to intensely hostile with physical acting out.

There is little in the record for the last week of November beyond the social work notation that Daughter B was the second and only surviving daughter. Mrs. Fisher wrote that Mrs. B "talked of wanting to die, her dead daughter who was an angel and was good," then referencing (Daughter B) "who was the bad daughter" (Fisher, 1954b).

The mother was leading the way in the closeness–distance cycles first observed at Menninger, showing up in both families here. Bowen wrote to a colleague about the project.

> I am attempting to hospitalize mothers in the same rooms with their schizo-
> phrenic daughters .... One gets a new respect for schizophrenia under such
> circumstances when you see a daughter gain her composure and attempt
> to pull away, only to see mother fold up. Then the daughter lets go of her
> goals only to fall back into the morass herself and have mother pull out. It
> takes some doing to get mother 'zeroed in' to the point that she is capable of
> risking psychosis herself in the interests of some gains for the two of them.
>
> *Bowen, 1954d, 2*

Bowen was using observations for staff education. In these early months, he devoted
significant time to staff training, staff observations and working through kinks in the
operations. He had daily clinical meetings with the two daughters.

The larger hospital system

Bowen made initial contacts in December with another family, referred to as the
Genain family. The broader administration brought them to NIMH in January.
This family, if part of Bowen's project, could have allowed for observations of two
parent–sibling interaction.

> I am completely intrigued by the possibility of an unusual addition to the
> project. The only set of living monozygotic quadruplets have been schizo-
> phrenic for some years and have about exhausted the possibilities of their
> present institution. It looks like I may have these four 24-year-old girls plus
> their mother and their father all housed in adjoining rooms.
>
> *Bowen, 1954c, 2*

Though Bowen's note implies that the family would be part of his project, they
were not directly part of his clinical study. NIMH had them come there for research
because of the uniqueness of quadruplets diagnosed with schizophrenia. Initially,
they stayed on Bowen's ward. Though the therapists for the sisters held many
meetings to discuss a research project, no specific investigation was in place for
months after their arrival. Bowen was the therapist for one sister and the family was
important for documenting development of the differentiation of self-scale and the
multi-generational family process. I discuss the Genain family in Chapter 15.

Family A, December 1954

Mrs. Fisher was on vacation for the last two weeks of the month. During her
limited availability for clinical sessions, Mrs. A revealed her anger toward the previous hospital, where basically she had been told to leave her daughter in the state's
care. This approach was not uncommon for acute hospitals. At the previous facility,
the doctor did not speak with her; she had only brief contacts with nurses; the
hospital restricted her visiting, and she felt criticized for the cause of her daughter's

problems. Mrs. A told Mrs. Fisher that she had easy access to the psychiatrist on Bowen's project (Fisher, 1954a). Social work notes also mention mother's upset when separated from her daughter.

The struggle to be a separate person took a physical form for Mrs. A. Taking an evening off for herself on December 1, she developed severe laryngitis. "The upper respiratory infection (December 8–10) coincided with Mrs. [A.'s] conflict and decision about leaving daughter for one week in early January to visit her sister in Ohio" (Fisher, 1954a). In attending to her physical problems, she was distracted from her daughter. The project's flexibility allowed Daughter A to have a pass to visit her ill, homebound mother. Nurses observed the daughter to be more age-appropriate when mother was focused elsewhere. The project's thinking was the mother's physical symptoms were not random, but were evidence of reactivity at the most basic level of the twosome. If either member of the pair could recognize this as a pattern in functioning, perhaps that understanding could help one of them make further changes in the relationship.

> In mid-December, about six weeks after admission, the mother was still at home convalescing from an afternoon trip. The patient became a little impatient with mother's illness. She asked if the staff would drive her into town to see mother. The staff did. That evening, mother had a severe gastroenteritis superimposed on her previous convalescent state. The patient began to ask if she could spend Christmas at home. Mother had dozens of questions. Did the staff advise it? Was the patient strong enough? Would the patient get upset?
>
> *Bowen et al., 1955, 11*

The two spent Christmas together. Next day, after spending Christmas together, the mother had a resistant infection (11). Mrs. Fisher captured the interchange at Christmas in her notes.

> The patient stayed at home with mother the entire Christmas period. The day after the patient returned, mother awakened with [an infection] … she said it did not have anything to do with the patient, it was caused by the raspberries she ate. Mother was treated by an allergist with little improvement …. She flew to another state for New Year's visit with relatives. She still had the (infection) when she boarded the lane. It had cleared before she arrived at her destination. She was 'snowbound' and stayed an extra week. The hives returned on the flight back. This somatic response to the patient was intense for two months. It then gradually subsided over about two more months.
>
> *11*

Family B, December 1954

In early December, Bowen discharged Mrs. B and gave the daughter a pass to return to their home city to settle issues related to where they had lived. On their return to the ward, Mrs. B's hyperactive behavior exhibited in drawing, writing and hypersocializing. The daughter did this behavior as well. "Together with [her daughter] Mrs. B. would run up and down the halls greeting everyone …" (Fisher, 1954b, 1).

The family's referring social worker visited the first week of December to give historical information to Bowen and Mrs. Fisher. Mrs. B met with the referring social worker and related how pleased she was to be at the Clinical Center. Then, in an illustration of what the project staff had to deal with beyond family interactions, Mrs. B asked the social worker to have any non-white staff fired. Bowen discussed such incidents in staff meetings, speculating that this behavior related to Mrs. B's early developmental adaptations to deprivation in her family. These were survival efforts. For her, the world was full of threats. Bowen suggested that Mrs. B, when understood in terms of her early life, did not appear so paranoid. Her history of evictions and conflicts with landlords and other tenants reflected patterns of fights within her family constellation, now projected on to society. She had an orientation to life that the world was "ungiving." This translated to her behavior of expecting to receive things for nothing (Fisher, 1954b, 1).

The staff created a treatment contract with her to offer choices to meet her unmet needs. The contract established rules within the hospital and within the ward for both staff and family member to maintain (1).

Both Bowen and Fisher considered Mrs. B too disturbed to go to the social worker's office. So sessions took place in Mrs. B's room five times from December 13 to the 18th (1). Daughter B sat in on these meetings. On December 15, Mrs. B attempted to have Mrs. Fisher assume financial responsibility for the family. Mrs. Fisher deflected this by supporting Mrs. B's own efforts, which produced an angry response by Mrs. B (1). The records contain many instances when staff maintained their own boundaries, gave support to "the adult" and did not shift to caretaking these poor, helpless people. Family B responded to this with intense negativity. The nurses and social worker set out what they would or would not do. Doing that elicited a flashback response. Staff observed this pattern often enough to consider it predictable reactivity in relationships.

Within a month, Mrs. B wrote of her complaints to Cholden, telling him, not Bowen, that they were considering leaving the project. Mrs. B cited the lack of attention from staff and the intercom system's tracking their conversation in the family's rooms and in the lounge. Families knew before admission that the intercom was part of the researcher's recording of family exchanges and that they could turn it off in their room.

> I am writing you a letter to tell you that they just simply leave us alone around here especially at night when we want to go to sleep early …. They also put the intercommunication system on in your office and if you don't believe me [a staff member] is a witness. I hope you all can rectify it, this situation at once. Cause we don't want to leave. [The referring doctor] will be very disappointed in us if we leave here. I hope you will rectify this at once.

> We just simply cannot stand it anymore. If it can't be rectified, we are going
> to leave for good. Cause we are really good and disgusted with what has been
> going on here lately.
>
> *Mother B, Name redacted, 1954*

In this letter, Mrs. B's wish to stay is obvious, but she had no sense of how to do that within herself. The letter shows functional helplessness, an effort to draw in someone else to solve her problem and to avoid finding a solution within self. Mrs. A got the problem off her chest and put it on Cholden's.

Family A, January 1955

Mrs. A, on her own initiative, visited her oldest daughter in another state the first week in January. Mother's usual vacillation was gone when making this decision. Daughter A requested and got passes to stay with her aunt over the weekend while mother was away. While visiting, she fit in to the family's life, caring for her young cousin, even baking a cake for her mother's birthday (Thompson, 1955). These records show the reverberations in functioning and support the idea of family as an emotional unit. In this month's report, the social worker noted the mother's efforts to manage her tilt to forced mothering, an important thread from the Menninger explorations. Mrs. Fisher described Mrs. A's concern over the daughter's whereabouts when out of her mother's sight. Mother saw only her daughter's dependence on her. Paradoxically, Mrs. A described the intense attachment as the daughter's problem "and that she herself could very well have some separation from [daughter] if [daughter] would allow it" (Fisher, 1955a). The remark highlights some awareness within the mother of the emotional process between her and her daughter.

The Genain family's admission on the 9th disturbed the ward's emotional balance. This is the family with quadruplet, schizophrenic daughters. The notes show reciprocal responses as Mrs. A claimed she was part of this disturbance, a point the social worker recognized as "overvaluing self." One variable of differentiation is a clear, realistic estimate of self. Although Bowen and the staff saw variation in functioning across the project's three families in these early months, the A family had the most adaptive flexibility. In a later discussion, Bowen recognized this family as having the ability to stay "relatively free of ward process and make good progress" (Bowen, n.d.).

Mr. A visited on January 17. Mrs. A arranged with staff for him to show slides of his work. He was a world-traveled lecturer. She invited friends to attend and stayed herself. Mrs. A said she did this to provide an audience for Mr. A and an enjoyable experience for her daughter. She disclaimed any possibility that it was related to her and her ex-husband (Bowen et al., 1955, 9). This differed from the father's earlier visits, when the mother was never present. Though she was his chauffeur on those visits, meeting him at the airport and driving him, she waited in the car while he visited their offspring. Changes in their daughter were significant by the end of December.

Beginning at Christmas, the patient spent every weekend with mother in town. Mother carefully arranged each weekend so another person would be with them. She frantically invited nurses or other patients to fill in when her friends were not available. In the meantime, the patient had become a gracious person in her social activities about town. She had obtained a volunteer job here at the Clinical Center and she was working several hours a day. She was seen 30 minutes daily by her psychotherapist. Mother was seen daily by the social worker.

Bowen et al., 1955

Near January's end, Mrs. A developed serious physical problems requiring hospitalization for a cervical biopsy. Her own mother had died of this cancer at the same age, facts she withheld from her physician. Her responses to scheduling the biopsy showed an intense reaction. She made excuses when canceling events. This included a planned trip to visit her other daughter again, which conflicted with the scheduled surgery. Uncertain how to manage her hospitalized daughter, she delayed telling her until the day before the procedure. By avoiding personal conversations with either daughter, she was avoiding the upset she expected in Daughter A. Once she told her, the daughter's response was perfectly appropriate. She maintained her stability through this time and visited her mother, bringing her flowers (7). Taking on an aunt's role with her young cousin, she and the niece went out for some pleasurable activity on the weekend that mother was in the hospital (Fisher, 1955a).

There are threads back to the regression explorations in the thinking of staff and how they observed Family A. Staff training on avoiding "forced mothering" behavior toward either mother or daughter transferred to at least one mother. The staff observed Mrs. A practicing restraint from giving directions to her daughter or correcting what the daughter did (Fisher, 1955a). Another observation also seen at Menninger was the effect of a "perceived demand from another." Mother had arranged for the head nurse to accompany the daughter on her visit home with mother for the weekend. When the nurse made a casual remark about the mother and daughter one day returning to their own home, Mrs. A's anxiety flared. Hearing this remark as her daughter sought a weekend pass, Mrs. A sought reassurance from both Mrs. Fisher and Bowen that her daughter was not ready for discharge. She made a point of impressing upon them the "seriousness" of her daughter's illness (Fisher, 1955a).

When Mrs. A engaged in an activity with her husband, declaring she was doing it for her daughter's benefit, this illustrates fused boundaries. By this time, the daughter had shown she could act for herself. These subtleties strengthened the idea of the family as a single emotional entity.

From November to January, Daughter A went through three phases: (1) tell me what to do, (2) a dawning awareness that she could make choices and (3) using her own initiative for what she wants. Mother reflected phases 1 and 2 in her functioning.

Family B, January 1955

January was a hard month for Family B. Mrs. B's ongoing psychosis meant the social worker had 28 sessions in the mother's room, sometimes twice a day. Five sessions took place in Mrs. Fisher's office when Mrs. B became more stable, beginning on the 26th. On the 9th of this month, staff admitted the Genain family to the ward. This was not one of Bowen's research families. The record shows a connection between this admission, increased disturbance in Mrs. B and the staff's negativity to her.

> The admission of the [Genain] girls on January 9, 1955 was another sig-nificant factor in the illness of Mrs. [B]. At the time of ... admission, Mrs. [B.] was overtly psychotic and she and daughter were extremely noisy. The [Genain] family complained about the [B] family and asked to be transferred to another unit. The [B] family was held primarily responsible by the staff for the leaving of the [Genains']. Mrs. [B] undoubtedly sensed this and no doubt entered into her continuous question of attempting to find out where she was to blame and her repetition of not knowing 'our rules and regulations'.
>
> *Fisher, 1955b, 2, 3, 4*

After the admission of the mother, father and four offspring—an intact family—Daughter B was aggressive toward her mother. The notes report that Mrs. B received serious beatings from her daughter, who pressured mother during these thrashings to correct her behavior. Mrs. Fisher described Mother B as not only accepting the assault but positioning herself to receive it and "invite it" (2, 3, 4). Observations led to questions: Is the daughter trying "forced mothering" on her own mother?; Or trying to force mother to a dominant position?; Is mother acting in a way that pushes the daughter to impairment?; If a young child lives with a psychotic mother, are there advantages to acting crazier than mother, or is Family B showing the intact family just how crazy and helpless two women are with no man to help?

The staff had qualms about continuing the treatment plan, including being a neutral presence, under the conditions of Mrs. B "inviting" physical aggression. Nurses considered restricting her to her room "lest she get out in the hall and invite somebody to start hitting her" (Bowen, 1955d). The staff had a hard time not inter-vening. By January's end, the staff interrupted the physical fights without taking sides. The rule was that "fighting was not allowed in the hospital" (Fisher, 1955b, 2, 3, 4). This decision, of course, set the staff up for repeated interventions with this family. If the staff tolerated such incidents and merely observed them, would either mother or daughter have ended the fights? These two had a long history of such incidents without serious harm before their admission here. The treatment effort was to always relate to the adult in the person. This was difficult for staff as there did not seem to be any adult present.

Mrs. Fisher took up her concerns regarding Mrs. B with Bowen:

> In a conference with Dr. Bowen .... I pointed out to him that it felt to me ... that I was talking to a small infant. Dr. Bowen agreed with this in part but suggested that .... I would not get too far away from her wish to be a mature person. He stated that Mrs. [B] probably will always have to have a supportive relationship and perhaps [Daughter B] will eventually be able to give this to her mother. 'Our optimal goal with Mrs. [B] would be for her to allow [Daughter B] independence from the intense attachment and for Mrs. [B] to form for herself a major dependency on the outside'.
>
> *2, 3, 4*

Change was possible in this family, given the right environmental conditions, even if the reality of the mother's capabilities limited the prognosis. The families represented different levels of functioning on the lowest quartile of a scale. An alternative to the use of an individual diagnosis was percolating in Bowen's mind.

Family C, January 1955

The first contact with the third research family, Family C,[4] came in October 1954 when Bowen visited the Women's Service at a local metropolitan hospital looking for prospective families. His schizophrenia/symbiosis criterion was an intense, overt attachment between parent and offspring and a diagnosis of schizophrenia in the daughter. There were several prospects, but the C duo had the most intense, enduring and vigorous attachment and influenced the direction of the future selection of families with fathers and siblings. "This experience made us cautious about working with further multiple sibling families until we can understand the process in small families" (Bowen, 1956b). The mother, age 42, and daughter, age 14, looked alike in their facial appearance, but were mirror opposites in physical size. Daughter C looked six to ten years older than she was (NIMH Clinical Record, 1955).

From October 1954 until January 1955, Bowen and the social worker held several interviews with Mrs. C. When offered a choice, the mother said she would stay at home, as she had two younger children. She promised full participation with frequent visits. Mrs. C labeled her divorced husband "an epileptic." She claimed his diagnosis was a part of the problem with at least two of her children, though neither had epilepsy (Basamania, 1955, 41). Mr. C had started a new family several thousand miles away and never visited his daughter at NIMH (Bowen, 1962). The criterion of the most intense symbiosis resulted in selecting mother–daughter pairs where fathers were no longer living in the family and the mother had no peer support.

Mrs. C's biological mother had died soon after Mrs. C's birth and the maternal grandmother raised her. She considered the grandmother her mother. Her biological father had left the family and remarried, with rare, brief contact between him and Mrs. C since then. Mrs. C's daughter arrived at NIMH on January 17, 1955, a few days after Daughter C's 15th birthday. Mrs. Fisher noted the "helpless" position of mother in a phone contact on the day of her daughter's admission. On the 17th, mother called Mrs. Fisher for instructions on what the hospital wanted her to do and asked for directions on how to get to the hospital (Fisher, 1955c).

The dance of together/apart was easily observable from the very first day. Once at the hospital, the daughter refused contact with her mother. Mrs. C then left to visit her grandmother out of town (Fisher, 1955c). Very early on with this family, Bowen thought about going up a generation and adding that maternal grandparent's generation to his research (Bowen, 1979, 3). This idea stimulated his early thinking on a sustained phenomenon crossing multiple generations.

On mother's first visit upon her return to the hospital, Daughter C pleaded for her to take her home. The mother ended the visit, went home and avoided saying no to the daughter. A consistent observation of the three mother–daughter pairs was an inability of the mothers to take a stand without invoking a voice of authority to support that stand. Mrs. A, for example, said, "Well the doctor/nurse/social worker/hospital says you can't do this." This shows that the mother's uncertainty contributed to the symbiosis. It was evidence that all were part of a larger process. Curiously, Daughter C's admission did not upset Mrs. B as had the earlier introduction of the Genain family. Daughters B and C, close in age, soon formed alliances with each other that depended on the status of the relationships with their own mothers.

Reading these notes, one sees a shifting hierarchy forming on the project, with Daughter C at the bottom. Her mother visited the least, and Daughter C was the most vulnerable to attacks from Family B. But when the two daughters aligned against Mrs. B, she then became the most at risk. At other times, Mrs. B and her daughter aligned against staff and Daughter C. This is a powerful example of how a third person can form many constellations of stability to two people.

The notes of both social worker and research assistant for January record the blurring of strength/helplessness in the three families. Mrs. A projected her own helplessness on her daughter and, at one point, gave a gift to Mrs. Fisher, signing it "Kindness of (Daughter A)" (Fisher, 1955b). Where does one person end and another begin? Family B had actual physical fights over who most needed help and who was in charge. The Cs lived out the closeness/distance cycles with each member inconsistent and upset tolerating each other's presence. An "I can't live with you" behavior was on display in Daughter C, telling mother not to visit and telling the nurses to inform mother of this. In return, the mother pressured the nurses to tell her how to respond. When staff told each of them that this is between the two of you to figure out, Mother C's attention shifted to her younger son at home, now having varying symptoms and needing her. Mother gave son's needs as the reason for her ambivalence about visiting her daughter. When mother was not visiting, the daughter exposed herself, was overly affectionate to staff or inserted pins in her arms.

Such acts showed to Mother, "I can't live without you." These acting out episodes occurred when Mother C did not visit and during periods of intense conflict between Mother B and Daughter B. The alliance between Daughters B and C was a way for each to manage self on the unit. Daughter C's act of pushing pins in her arm occurred after hearing of mother's threat to remove her from the project. This threat is a version of forced mothering toward Mother C as well as a warning to the research program: Do something or I'll mess up your investigation. Mrs. B also used this tactic in December.

When the social worker from the referring hospital visited Mrs. C on the 31st, she was full of complaints: Her daughter was doing worse at NIMH and there was no help here for her. The previous hospital would have told the mother how to cope with this. Mother did not see in herself the ability to think for herself, to find within a solution. The answers were with others. Nor did Mrs. C see that her daughter's acting out was to her mother's agency. Mrs. C pleaded with her previous social worker to tell her what to do.

Bowen wrote of this family:

> The third pair came after Christmas. It had taken her 2 months to get herself worked to the point of participating while her daughter stayed in cold packs at St. E's. This mother has reared 5 children on welfare money without a husband … the pt. who is 15-big and hefty and well developed. The first visit here, the mother went to the gym with the p[atien]t. The pt. participated little, but the mother played soft ball, pitched basketball baskets, and did a 25 minute workout on the trampoline alone. Two days later, she wanted to bring the two younger boys to spend the day here with her. We held the line of feeding the kids and she now leaves them home, but if we would permit, I am sure we could soon have the mother, all five kids, son-in-law and a grandchild. As is, she simply cannot allow herself to regress because she has to go home nights to the others.
>
> *Bowen, 1955e, 2–3*

The helplessness seen in all three mothers prompted the question, "Who has the problem?" Sometimes the three daughters had higher functioning than their mothers, but they could not sustain it. Only one family member could be strong at a time.

Theory develops

When staff did not comply with the pressure by family members to "tell me what to do," the family members were negative toward staff. Recognizing this posturing as being reactive to the growth opportunities helped staff to keep their neutrality as best they could. These observations contributed to the new understanding that these mother–daughter dyads were fragments of a larger relationship formation. Each of the three families had previously depended on a larger system: a hospital system with Family A and Family C, and a social welfare system with Family B. Every visit by someone outside the project was an event for the staff to analyze and learn from. When a father visited, shifts in the family member's relationships were visible. It increased the staff's awareness that more people could add strength, resources and flexibility to a family's choices, a construct of the family functioning as a unit.

As discussed, the three families showed "I can't live with you, I can't live without you" closeness/distance cycles in vivo. Too much togetherness brought rejection. Distancing is an effort to hold on to self in a relationship. This idea remains an important one in Bowen theory. Rejection looks the opposite of what it is. Bowen regarded it as a reaction to the intense connection between people where one person gives up self to the other and then pulls away to find self again. This observation goes back to Bowen's Menninger explorations and came from the intimacy-distance cycles observed in the relationships there. He planned for it in studying relationships at NIMH. It is a distinct understanding of a natural system. Living things automatically space themselves in relation to another. In humans, this includes emotional space.

The shifting dependencies in the B and C families could be choreographed. In both families, mothers and daughters demanded actual parenting from the clinicians and were expert at finding caregivers among the personnel. Daughter C, in the rejection cycle, turned to the nurses or aides or to Daughter B for her mothering needs. Daughter B formed a loving attachment to the nurses or to the daughter in Family C. This produced extreme negativity and threatening behavior in Mrs. B toward Daughter C, and often she acted out with threats displaced to the staff. When Daughter B was not available to Daughter C, Mother and Daughter B allied, presenting intense hostility to the staff as a pair. As soon as the two were in lockstep, closeness came easily. This family was one of the inspirations to start the family-staff group in August 1955, which then became family psychotherapy in the second year of the project.

Mother C, shortly after rejection from her daughter, turned to her 14-year-old son, who developed severe anxiety symptoms. "One mother could separate from the hospitalized daughter and within hours her anxiety would be less and a younger son would have disabling neurotic symptoms" (Bowen, 1957, 9). In the three families, turning to another person (a sister, staff, another daughter on the ward, a son at home) was a method to manage the person to person, mother–daughter relationship. Two people could not manage the disquiet between them without involving a third person. These are historic points in the later conceptual development of the theory.

Family A, February 1955

Fluctuations in functioning within the mother/daughter relationship continued. Fifteen times Mrs. A and her daughter were out on pass together. Daughter A was improving. Early in the month, on a weekend visit Mrs. A had a migraine headache and the household help was off duty, so her daughter took over. This was two days after Bowen had sent a report to Daughter A's previous therapist summarizing her progress. Bowen wrote of the rapid progress made at NIMH, describing how she had gone from a regressed state to "one of a charming and gracious young lady" (Bowen, 1955e). He summarized her volunteer work at the hospital, participation in intra/extramural activities and competence to have day and weekend passes with her mother at her aunt's house. While the fundamental problem remained, her functioning was much improved. Bowen referenced the mother's presence as "a catalytic agent" toward change (Bowen, 1955e). In receiving praise, Mrs. A again feared that she and her daughter were being considered for discharge. If discharge were pending, the mother, based on her actions during the weekends, was not ready to live with daughter, and that presented problems for both of them. Neither mother nor daughter had within self the strength to regulate, without symptoms, the closeness/distance in their relationship (Fisher, 1955d). The daughter did well with the hospital's structure, including as much contact with mother as each could tolerate. For mother, that was far different from living together in an unstructured environment with the potential for serious disturbance returning. There were many observations of excessive closeness and over-distance. Mrs. Fisher noted that mother attributed responses to her daughter though the responses were about herself. Whatever Daughter A did, Mrs. A measured it by another's approval of herself. Each of the women depended on the hospital structure to regulate their tolerance and comfort in each other's presence.

The social work notes, in contrast, related that the mother considered taking her daughter home for the summer. Bowen noted mid-month that Mrs. A had told him she would consider living on the project "if I wanted her to." Bowen questioned if this had to do with the Clinical Center being air-conditioned (Bowen, 1955e, 2).

> Now on to mechanics of the mother-patient business, I have three pairs. The first was a 31 yr. old schizophrenic woman who had been psychotic since 15 with 17 years of in and outpatient treatment. She had over 12 years of 'psychoanalysis.' My proposition was that I felt the presence of mother would benefit treatment and we're setting it up so mother could spend a great deal of time here and even live right on the ward. This mother chose to live out but to come in very day and .... Work ... with the patient in OT. After 4 months she has said to me that she would come in if I wanted her to. She has admired our air condition as a retreat next summer. She has resisted the patient's numerous attempts to have her spend the night.
>
> 2

For Mrs. A, pleasing authority figures was natural. Deciding for herself to increase proximity to her daughter was another matter. The nurses wrote that there was very poor direct communication between the two. And Mrs. Fisher's notes showed mother had many private wishes for her daughter's improvement. Mother thought of her daughter as if she were a teenager, not a woman in her 30s (Thompson, 1955). Both sought nurses' advice to navigate the environment and deal with conflicts between them. Each asked staff to use an order or a hospital rule to set the tone of their relationship.

They do not have the capacity within themselves to maintain the relationship at a comfortable 'not too close and not too distant' level. They reach out to have the environment make a rule, give some advice, or invoke a hospital regulation that controls distance and prevent each from being exposed to the intensity of the conflicting wishes. So great is their wish for structure that they will use a chance remark from a staff person as a basis for a rule ascribed to the hospital.

Thompson, 1955

This point made staff cautious on complimenting Daughter A's improvement and they found that a neutral comment caused less rise in anxiety than any praise did.

Family B, February 1955

Mrs. B was too disturbed to leave the ward this month. She saw the social worker in her room. On the 11th, Mrs. B made threats to leave, followed by a demand that someone drive her to Baltimore (Fisher, 1955e). Bowen wrote that Mrs. B's actions were related to her background. Such behavior had generated problems wherever she and her daughter had lived. He wrote that they "had literally lived in dozens of welfare apartments always moving over some paranoid fight by the mother" (Bowen, 1955e, 2). As noted before, Mrs. B regressed on admission in November, and by February "she has been as psychotic as a patient can get. This mother is the most impaired of all my six [that is, the three sets of mothers and daughters]" (2).

The project considered the mother to be the adult in the family, better able to handle life. Nurses looked for evidence of adult behavior as an entry point for giving them support. Living that out was difficult for these mothers.

This month, Bowen observed the first evidence of counter transference between a staff member, Mrs. Fisher, and a family member, Mrs. B. It was a learning opportunity for everyone to consider interlocking triads. Here, countertransference meant that Mrs. Fisher's feelings toward Mrs. B colored her analysis of reality and her responses to Mrs. B in the setting. "The feelings of empathy and identification are very much a part of the phenomenon of countertransference ..." (Bowen, n.<u>d.</u>, 16). Bowen described this as a growing protectiveness of Mrs. Fisher for Mrs. B. Mrs. Fisher was defending Mrs. B in meetings and had concerns about her depression. None of the other staff saw this (Bowen, 1955e, 1). Mrs. B had expressed only to Mrs. Fisher that she wanted to leave. She was having great difficulty living in the hospital (1). Secrets or confidences such as this, while contributing to the countertransference here, could skew the data and cause problems for the research. Bowen later addressed it by establishing a principle of openness.

Prompts for the countertransference came from outside too. In violation of the agreement between both the B family and the research project, the referring agency had ended the B family's lease. They removed the family's belongings from their apartment without informing either Mrs. B or Bowen or Fisher (1). Mrs. Fisher had assured Mrs. B that her apartment was there if she wanted to return to it (Bowen,

1955e, 2). Angry at what the agency "did" to Mrs. B, Mrs. Fisher called the social worker about this disregard for both Mrs. B and the project (2). These two social workers were friends. Open commentary was acceptable between them, but Bowen saw this incident between the social workers as the surface tension of the counter-transference: Mrs. Fisher's wish to take care of Mrs. B.

The broader context was the social worker's increased motherliness seen in her protective attitude toward Mrs. B. and with the referring agency. Another incident showing how all parts of an organization contribute was Mrs. B's demand that Bowen get Daughter C to behave. He took a non-interfering position in responding to Mrs. B. It was up to the Bs

> to get along with the situation the best they can. Implicit in this stand I took but not verbalized at this point, of course, was the stand that I also would make no move to make the B family behave when they are pestering Miss C.
>
> *Bowen, 1955e, 2*

Bowen explained this process at the staff meeting. The emotional source was the alignment of feeling between Mrs. Fisher with Mrs. B, the negativity to the Daughter C and the staff's perception they had some responsibility "to protect Mrs. B" (2). It shows that an extrusion process was operating toward Daughter C, while Mrs. C's absence also contributed. And it shows how difficult it was for staff to live out the effort of being both a resource and an observer.

The unfolding of events portended a revised hypothesis.

Family C, February 1955

At the start of February, and after surgery to remove a bobby pin in her arm, Daughter C regressed to where the nurses took care of her as if she were an infant. Bowen understood this as an opportunity to give her mothering that prevented acting out behavior. At first, the nurses had a parental reaction to "'straighten up and fly right' or 'act your age'" (Bowen, 1955d). Bowen's experience at Menninger helped him direct the nurses to work hard to find their way to giving without judgment. These episodes illustrate the complexity of following a theoretical path challenged by real-life experiences.

February found Mrs. C in financial distress. She had 12 sessions with Mrs. Fisher, but money problems caused her to miss her appointments the last eight days of the month. The daughter refused involvement with her mom. She sought excessive attention from the nurses, while the mother lamented to the social worker that she had lacked her own mothering growing up (Fisher, 1955c). She seemed to feel angry with Mrs. Fisher as well.

> [Mrs. C.] was expressing this in many ways, including one statement that it would be easier for her to be the social worker and Mrs. Fisher to be the

patient and then she could be quiet and Mrs. Fisher would have to do all the talking. Mrs. C. was expressing quite a few angry feelings to a number of different people about Mrs. Fisher.

Bowen, 1955e, 3

These observations in the staff meeting discussions led to a suggestion that Mrs. Fisher offer more to Mother C. Describing Mrs. Fisher's implementation of this, Bowen wrote:

After this, she began to lean over backwards to be more outgoing and giving to Mrs. [C] This resulted in a rather beautiful piece of acting out by Mrs. [C] two days later. This was on Monday, February 7. Mrs. [C] arrived for a 10 to 10:30 am. appt. after 10:30. Mrs. Fisher was at that time seeing Mrs. [B.] Mrs. Fisher went to the door and said that she was busy now but that she could see Mrs. [C] at 11 a.m. Mrs. [C] said she would be there. Later we heard from the nurses that after Mrs. [C] got on the ward, she decided she did not want to see Mrs. Fisher and the appointment was canceled. At this point the sibling rivalry between these two mothers, Mrs. [B] and Mrs. [C], which had been smoldering, now broke out into an open expression of ill-feeling.

Bowen, 1955e, 4

Countertransference from Mrs. Fisher toward Mother B emerged after the C family came on the unit. Bowen's assessment was that both Mothers A and B were vying for favorite child position (4) with Mrs. Fisher. Acting in agreement with the old saying that actions speak louder than words, Mrs. C then refused to meet with Mrs. Fisher. Mrs. B then followed up by threatening to leave and "mess up the research." Was it Mrs. A, B or C that the social worker cared for most?

Bowen addressed Daughter C's earlier acting out as a continuance of what existed between mother and daughter before they entered the hospital. It was how the two operated with each other. By the end of February, Daughter C asked to go in seclusion to gain control of herself. By granting this, the nurses were offering the daughter support for her own initiative. Bowen chose this family as a lesson in the symbiosis's externalization. The plan was to create

a ward environment which could give the support and not get pulled into the 'middle man' position of siding with the patient to make the mother's behavior according to the 'patient's wishes nor to side with the mother to make the patient behave as the mother would have it …. Before we started, we realized that this would involve … intense and very emotional 'family fights' in which both mother and daughter were adept at getting partisan support for their own viewpoints. We proposed to be able to give support to each and to side with neither.

Bowen, 1955e, 3

As Mrs. Fisher gave more attention and time to Mrs. C, upset flared between daughters B and C. Daughter C shifted back toward mother, going on a home visit with her. Reciprocally, Mother C turned her attention to the son at home. Daughter C, once back at the hospital, got into conflict with Mrs. B. If one thinks of this as a dance, it involved moving close, changing partners, twirling and dramatic interchanges. These shifts were dramatic as Daughter C had suicide ideation and self-mutilating behavior through these days related to the substance of the mother–child interactions. Nurses on 3-East had committed to a belief that every person can grow given the right conditions. They were committed to applying the idea that change is possible in severely impaired people, moreover that change from within a family lasts longer. It did not always work well.

Problems with the research also arose from the broader context of hospital setting. Bowen's research project was set in half a ward where nurses carried out very different protocols from those of the doctor and patients on the other half of the ward. Bowen wrote of this about Daughter C. "From January 1955 until March 1955, ward staff was divided between two projects and intrastaff differences were 'acted out' between the two projects" (Bowen, 1955a, 3).

Some incidents came to the administration's attention

The families certainly tested the nurses' and aides' commitment to principles. Here is an example of the gravity of one challenge presented to the nurses.

> My ward is not the usual ward here. One of the most interesting pairs is a 15 y/o very buxom and big acting out girl almost of the acting out potential of little Mrs. [name redacted] at the clinic. Her mother a mirror image opposite. One day this masochistic girl 'surprised' my nurse by coming out with safety pins stuck all the way into her forearms. The nurse acted like she didn't even see them. The girl asked to go to OT- [in another part of the building]. The nurse told her she'd have to take out the pins if she went out. Last Sunday (Feb. 13) afternoon came the pay off. The patient had been talking about suicide, which was a not a too infrequent threat with her. She came charging out of her room with a belt noosed around her neck and barricaded herself in the bathroom. That was 1:30 pm. At 4:30, when the patient sneaked back to her room, the nurse asked if she would like something to eat. Someone asked the nurse what she did when the patient did that. She replied, 'I went into the nursing office and had an anxiety attack' …. I heard about it in a routine report on Monday. This kind of nursing service is hard to come by but it is getting to be more the usual than the unusual on my ward. Nurses can develop some amazing capacities if they can get support in feeling of the situation and in backing their own judgment.
>
> *Bowen, 1955e, 1*

Morris Parloff, a psychologist contemporary of Bowen's at NIMH, discussed the hospital's response to such incidents in an oral history, in 2002. He studied the

impact of ward milieu on nursing staff. Bowen's project was one representation. Parloff noted that the hospital evaluated each incident of patient care. He
described another incident where a patient on the research ward developed cel-
lulitis and the Center brought it before the medical board (Morris Parloff, 2002).
Parloff's study addressed the challenges of these experimental environments for
nurses at NIMH.

> The milieu had an even more obvious and powerful impact on the ward
> nurses. This was due to the fact that the nurses … hired by the independent
> Central Nursing Department for their demonstrated mastery of basic psy-
> chiatric nursing skills, now found themselves having to work for psychiatrist
> ward administrators …. In short, these nurses were required to meet dual
> loyalties. They were to be loyal to the nursing standards set by the Central
> Nursing Department and simultaneously be loyal to the idiosyncratic nursing
> requirements demanded by their ward administrators …. Each ward adminis-
> trator had been hired by Bob Cohen on the basis of his putative creativity …
> most clearly evidenced by his unorthodox ideas. The ward nurses … found
> themselves having to adapt to the unique and sometimes bizarre requirements
> imposed by such psychiatrist ward administrators as Lou Cholden, Charlie
> Savage, Jordan Scher and Murray Bowen. Each of these ward researcher
> administrators was quite capable of driving even the most stolid and sober
> nurse—how shall I say it—nuts.
>
> *Morris Parloff, 2002*

Beyond their loyalties to Central Nursing and the research project directors, the
nurses had to meet administrative criteria. The Clinical Center had policies for
meeting basic standards of excellent health care. NIMH was government funded,
and it had to practice within very specific government guidelines while being
expansive in its approach. The research produced there balanced individuality and
cooperativeness at the highest level.

Thinking family

The original question about specific outcomes in offspring raised by the same
parents was still on Bowen's mind in early 1955. He wrote to a colleague,

> If you should run into a family with one schizophrenic child and another
> alcoholic child and it would be possible for both parents and both patients to
> come and stay, then I most certainly would want to hear about that.
>
> *Bowen, 1955f, 2*

The reference to having a complete family on the project portended the imminent
birth of the catalytic concept of the family as an emotional unit and considerations of a treatment that would match this understanding. Bowen's mind already
was there.

I have been itching to try a different kind of ward communication system. For instance, one could begin by never having a meeting between staff members in which patients were not invited or welcome. This would put some kind of strain on patients and staff alike, but I'll bet my group is up to it. Hell, I'd even be willing to make the nursing office and all the records wide open to everyone. Maybe even ask the patients to write their own records. My gang wants to try it, but if we started something as unusual as that, then how in hell would we know whether it was the presence of mother or us that brought about changes.

Bowen, 1955e, 2

The first inklings of initiating a change in the project toward family psychotherapy took place in February 1955. Prior to that, the emphasis was on the mother's presence. The hypothesis and treatment plan in place still looked like the family effort at Menninger, except for the greater involvement of the mothers and the 24-hour observations.

Because the mothers were not all living in, though, they were "not required to function as mothers" (Bowen, 1965b, 132). This often left the patients to nurses who worked particularly with one family. The larger field of evidence at NIMH and observations of the shifting attachments to staff during mother–daughter disruptions and fathers' visits were the essence of theory building. Bowen had the visionary's capability of creatively generating what was unprecedented. He recognized that "Rules and advice make for a calm ward but the pattern of the symbiosis is lost" (Bowen, 1956a, 8). Chaos on the project was a clue to pay attention. Something important was happening. In that disorder, he found a missing piece in his theory. Standing back, the movements of mother–daughter pairs revealed that a family meant more than one parent and one child. The smallest molecule of an engaged human system is more than two people. The mother–daughter pairs were only a fragment of a larger unit, an epiphany which came from the collage of observations.

Bowen brilliantly integrated all these observations to advance a new theory of the human. Introducing the idea of the family as a unit reorganized the entire trajectory of the research. Bowen now could begin to build a project that ultimately led to admitting an intact family in December 1955. It involved planning a structure for an idea that had no precedent, pursuing a treatment to match that idea and making presentations to others about the process of discovery.

As recently as 2017, Sheppard and Suddaby, academics interested in the construction of knowledge, identified what Bowen encountered as crucial to theory building: "an unexplained puzzle resulting from an unexplained phenomenon that defies extant knowledge" (Sheppard & Suddaby, 2017, 64). He filled in a missing piece of the theory within the first five months of observations. That piece would serve as the catalyst for building a new theory. Interaction between two family members had not been adequate to explain how the research families behaved. A hypothesis that fit these facts was now needed. By February, the project was clearly moving toward studying intact families with an altered hypothesis.

The records from these months do not have the confidence expressed here. Instead, records show a steady move forward, reporting on findings in a way that allowed for continuing exploration without arousing controversy. It is only when looking at the aggregate of those writings and the actions that went along with them that it is possible to see Bowen's confidence.

Notes

1 Families are named in alphabetical order of their arrival at NIMH. The first family is A, the second family is B and so on.
2 For this chapter only, I will review nursing notes (also known as patient charts or unit reports) of the initial three families from November 1954 to February 1955. It was then when Bowen tested his resolve to make public his observations of these families and his conclusion that the family was a unit. He put this into his March report to administration. The remaining months of 1955 were important for further confirming the concept of family theory. Records exist on notes of each family's interactions, but I will not review them here.

   The discussion that follows comes from charts by the social analyst, social work records, the research assistant's notes, a presentation to NIMH staff by Dr. Bowen, the head nurse and the social worker in August 1955, a narrative summary by Warren Brodey in 1957 and the unit report for February 1955; Unit Reports from November 1954, December 1954 and January 1955 have deteriorated with time and no longer exist.
3 For in-depth discussion of this family on the ward in August 1955 refer to: Rakow, C., "Learning from the nurses' notes for Bowen's 1954-1959 NIMH project: A window into the development of theory," *Family Systems: A Journal of Natural Systems Thinking in Psychiatry and the Sciences*, Spring 2016, Vol. 11, No. 2.
4 This mother daughter family were the subject of the paper "The 'action dialogue' in an intense relationship: A study of a schizophrenic girl and her mother" by Robert H. Dysinger presented at the Annual Meeting of the American Psychiatric Association, Chicago, IL, May 15, 1957 and published in *Family Systems Journal*, Vol. 6, No. 2, pp. 117–120.

References*

*Unless otherwise specified, the works of Murray Bowen included in this chapter are from the Murray Bowen Papers. 1951–2004. Located in: Modern Manuscripts Collection, History of Medicine Division, National Library of Medicine, Bethesda, MD. The Accession, Box and Folder information differs for each one and that is included here.

Basamania, B. (1955). [Clinical record, case work with (name redacted) Mrs. C. during November 1955]. L. Murray Bowen papers, National Library of Medicine, History of Medicine Division, (Acc. 2006-003, Box 3, Folder Clin. record copies), Bethesda, MD.
Bowen, M. (n.d.). [Draft. Family study clinical research project]. (Acc. 2007-073, Box 3, Folder Working papers from NIMH project re: schizophrenia).
Bowen, M. (1954a). [Letter to a Menninger patient on September 16, 1954]. (Acc. 2006-003, Box 4, Folder Personal correspondence).
Bowen, M. (1954b). [Letter from a Menninger patient's mother on September 21, 1954]. (Acc. 2006-003, Box 4, Folder Personal correspondence).
Bowen, M., 1954c [Analysis of NIH program activities project description sheet, December 1954]. (Acc. 2006-003, Box 4, Folder NIMH–annual report (project description).

Bowen, M. (1954d. [Letter to a Menninger colleague on December 30, 1954]. (Acc. 2006-003, Box 4, Folder Correspondence – personal).

Bowen, M. (1955a). [Memo to Robert Cohen on October 27, 1955, "Report of the clinical course of Miss (C)"]. (Acc. 2006-003, Box 3, Folder name redacted, Daughter C).

Bowen, M. (1955b). [Letter to Daughter A.'s previous therapist on February 3, 1955]. (Acc. 2006-003, Box 6, Folder NIMH correspondence H through P).

Bowen, M. (1955c). [Letter to Daughter A on August 31, 1955]. (Acc. 2006-003, Box 6, Folder NIMH correspondence H through P).

Bowen, M. (1955d). [Report on 3 East Project February 11, 1955]. (Acc. 2006-003, Box 3, Folder name redacted B family).

Bowen, M. (1955e). [Letter to Walter Menninger on February 16, 1955]. (Acc. 2006-003, Box 8, Folder Walter Menninger (Will's son)).

Bowen, M. (1955f). [Letter to resident under Bowen at Menninger on February 9, 1955]. (Acc. 2006-003, Box 6, Folder NIMH correspondence a through G).

Bowen, M. (1956a). [Transcript of Dr. Bowen's participation in combined clinical staff mtg. on March 29, 1956]. (Acc. 2006-003, Box 4, Folder Transc. of combined clinical stf).

Bowen, M. (1956b). [Formulation of 3 east family study project July 16, 1956]. (Acc. 2006, Box 5, Folder Project: Special seminar on schizophrenia).

Bowen, M. (1957). Treatment of family groups with a schizophrenic member. Chapter 1 in *Family therapy in clinical practice*. New York: Jason Aronson.

Bowen, M. (1960). A family concept of schizophrenia. Chapter 4 in *Family therapy in clinical practice*. New York: Jason Aronson.

Bowen, M. (1962). [Draft]. (Acc. 2014, Box 6, Folder Chapter I the research project).

Bowen, M. (1965a). [Letter to James E. Banta on March 15, 1965]. (Acc. 2003-026, Box 7, Folder Banta).

Bowen, M. (1965b). Family psychotherapy with schizophrenia in the hospital and private practice. Chapter 8 in *Family therapy in clinical practice*. Northvale, NJ: Jason Aronson Inc., 1978.

Bowen, M. (1979). Simon, R., "A network genealogy: A history of family therapy in the Baltimore-Washington corridor, Part II – Adventures in 'the great democracy': An interview with Murray Bowen". *Family Therapy Practice Network Newsletter, III*(5), 1–7.

Bowen, M. (1986). [Draft]. (Acc. 2006-003, Box 8, Folder Drafts of odyssey—in theoretical principle).

Bowen, M., Fisher, T., & Bowe, M. (1955). [Presentation to staff on August 19, 1955]. (Acc. 2006-003, Box 3, Folder name redacted Daughter A).

Brodey, W. (1957). [Clinical record narrative summary (A. family), on June 18, 1957]. (Acc. 2006-003, Box 3, name redacted Family A).

Family Services Social Worker, name redacted. (1959). [Case committee presentation by (name redacted Social Worker) on April 10, 1959], Family & Children Agency, East Coast, United States.

Fisher, T. (1954a). [Clinical record research record. Summary of casework help to Mrs. (A.) during the month of November 1954 through August 1955]. (Acc. 2006-003, Box 3, November 1955 (Folder name redacted Daughter A)).

Fisher, T. (1954b). [Summary of casework help to Mrs. (B.) during the month of December 1954]. (Acc. 2006-003, Box 3, Folder Clinical record copies).

Fisher, T. (1955a). [Summary of casework help to Mrs. (A.) during the month of January 1955]. (Acc. 2006-003, Box 3, Folder name redacted Daughter A).

Fisher, T. (1955b). [Summary of casework help to Mrs. (B.) during the month of January 1955]. (Acc. 2006-003, Box 3, Folder Clinical record copies).

Fisher, T. (1955c). [Clinical record summary of interviews during the months of January, February and March, 1955]. (Acc. 2006-003, Box 3, Folder name redacted).

Fisher, T. (1955d). [Summary of casework help to Mrs. (A.) during the month of February 1955]. (Acc. 2006-003, Box 3, Folder Clinical record copies).

Fisher, T. (1955e). [Summary of casework help to Mrs. (B.) during the month of February 1955]. (Acc. 2006-003, Box 3, Folder Clinical record copies), Bethesda, MD.

Mother B, name redacted. (1954). [Letter to Louis Cholden from Mrs. B. on December 1, 1954]. (Acc. 2006-003, Box 3, Folder name redacted B. Family).

NIMH Clinical Record. (1955). [3-E Unit report]. (Acc. 2006-003, Box 1, Folder, February 1955).

Parloff, M. (2002). *Dr. Morris Parloff oral history 2002 a*. National Institute of Health. Retrieved July 30, 2020, from https://history.nih.gov/display/history/Parloff%2C+Mor ris+2002+A.

Shepherd, D., & Suddaby, R. (2017). Theory building: A review and integration. *Journal of Management, 43*(1), 59–86.

Thompson, C. (1955). [Chart]. (Acc. 2006-003, Box 4, Folder Clare's meeting notes & research charts).

13

THE FAMILY AS A UNIT

March to July 1955

> My effort began in 1949 with individual psychotherapy for multiple members
> of the same family .... I think I was 'blind' to the total family relationship
> system with the early families because of training in individual theory which
> did not account for the family relationship system.
>
> *Bowen, 1965, 7*

Bowen quietly announced his seminal understanding of the family as a unit in a
quarterly report in March 1955. To give context to Bowen's efforts, this chapter will
introduce people and places of interest and importance in this period.

Preparing for a first quarter report

An unfinished essay compared and contrasted a "clinical psychotherapeutic
approach ... with the approach of science" (No Author, 1955, 1), dated one week
before the report presented at the March 15, 1955 staff meeting, asserts that the
"criteria for accuracy and credibility are the same for both" observations in nature
and in clinical settings (No Author, 1955, 8). Any clinician interested in "... useful
and reliable knowledge" [underline in original] must have tolerance for uncertainty". As "the approach of science is the approach which offers the way toward
solid contributions ... [t]he scientist thrives on mystery and is under no immediate
pressure to dissipate it to take action in a practical situation" (9). In contrast, "the
clinician needs to feel that he knows something about what is going on and what
he is doing and the presence of vast mystery is often difficult to tolerate" (9).

The assertions in this essay are compatible with Bowen's practice of free association, putting thoughts on paper to clarify his thinking. He now faced a decision.
His recent work suggested a new possibility, the family as an interdependent unit of
functioning. He could follow that reasoning or try to interpret it into an individual

DOI: 10.4324/9781003027287-14

theory frame of reference. Either was a challenge. He went with the data. "The first efforts to perceive the family as a unit began as a clinical necessity" (Bowen, n.d.a).

The past five months of observing interactions between mother, offspring and staff showed it was difficult to perceive them as separate individuals. Mother and daughter mirrored each other in "every detail of the patient's psychosis" (Bowen, n.d.a), conveying a broader context of action than an "interlocking of individual psychologies" (Bowen, n.d.a). Neither mother nor daughter had inner control over this reactive dependency. It could transfer and involve the staff, another child at home or another patient. The twosome could not contain the intensity of their interaction without getting someone else involved. Adding a third person gave both relief and balance and seemed contagious, adding others in to an interlocking extension. "The involvement of a single key person, such as the head nurse, could quickly spread by a series of administrative and interpersonal relationships until a good portion of the staff could fight the symbiotic battle" (Bowen, n.d.a).

All these were new understandings. Observations and reasoning formed the substance of the idea that the mother and daughter operated as a fragment of a larger emotional grouping. "Family" went beyond two individuals. It was a naturally formed grouping having an interactional phenomenon that regulated emotional intensity. These observations were not in the literature, nor were they part of Bowen's training. He had set up his project to find clues to explain human behavior and had the needed objectivity to see them. The repeated cycles and involvements of family members were consistent enough to be called patterns. These were automatic behaviors, occurring spontaneously, given certain conditions. Such phenomena met the project's purposes of going beyond common knowledge.

The family, as a unit, makes its debut

The first public reference to the family as a unit, on March 15, 1955, marks when Bowen put the idea into words and presented it to National Institute of Mental Health (NIMH). Bowen carefully worded this important discovery, couching his seminal finding in qualifiers. With a hint of a need for an alternative method of therapy, he paid respect to psychoanalytic psychotherapy. As a principle, Bowen did not use the word "believe" casually. Following this staff meeting, he moved toward a changed hypothesis and treatment plan, and the major step of adding fathers to the project dynamic.

> It is much too early to have any definite ideas about this, but it would appear that the impairment of the mother is of about equal degree with the patient's illness, except that it is manifest in a much different way. Observational material is rich. One of our greatest pleasure has been to see a rather rapid therapeutic response, which we believe to be much greater than would ordinarily be obtained with either or both in separate therapy. This conceptual framework is presented, not as a definite classification, but as a way of conceptualizing the various approaches. The basic one-to-one relationship

of psychotherapy has long been the basic unit of work in this analytic area, and its importance is great. It probably will continue to be the basic unit, especially in patients able to be treated in an out-patient setting. We believe that study of the family unit has a great promise, especially in research work and the understanding of problems. It may well have great importance in treatment of more impaired patients.

Bowen, 1955a, 6

In the next four months, Bowen sought ways to substantiate or to refute this revolutionary idea of the family unit. He used the term in the NIMH research project to convey the family as a single organism with reciprocating interactions and responses. Henry Richardson, "the father of modern family medicine" (Roy 1988, 9), used the phrase "the family as a unit" in his 1945 book *Patients Have Families.*[1] Reviews of Bowen's notes show that his usage of "family unit" began without knowledge of Richardson's book. That came later, when he found it during a literature review (Bowen, n.d.b).

Both Richardson and Bowen considered the family a biological whole. According to Monika Baege, who reviewed Richardson's book, Richardson reasoned that "the outstanding characteristic of the family, as a biological organism, is that it is composed of living members and therefore must change" (Baege, 2006, 181). Bowen reasoned a family is a biological entity shaped by evolution, like other living systems. Richardson studied one family at a time. Bowen studied multiple families across varying levels of functioning, from severely impaired to normal.

Bowen's understanding of the context of differentiation of self versus diagnosis changed. One family member had the diagnosis. Research findings suggested the problem was in both members. Reactivity in a family member could transfer to another. Two people were not enough to contain the emotional responsiveness. A third person was important to this stimulation. With the project's approach, therapeutic improvement was greater in mother and daughter than with each being in therapy separately.

After giving notice to the administration in the quarterly report, Bowen started a months-long search for what treatment method to use with a family. Bowen was a careful researcher operating in a clinical center. It was an ideal environment to fully consider the family as a natural organism. The new schizophrenic concept involving family was acceptable to the research facility, giving Bowen the leeway needed to take in other family members for the research project. He could maintain the emphasis on extending current understandings of schizophrenia and families without torpedoing his goal of a more factual theory.

Concurrent observations of others

Other professionals in psychiatry were paying attention to staff/family interactions and finding similar patterns to Bowen's findings. Warren Brodey and Marjorie Hayden, S. M. presented the paper, "Intrateam Reactions: Their Relation to the Conflicts of the Family in Treatment," at the March 1955 annual American Orthopsychiatric Association meeting. In this study, Brodey gave case examples of conflict among professionals working as a team with members of the same family, describing the intrateam conflicts as "a reduplication in diminished form of the family conflict" (Brodey & Hayden, 1955). He was seeing a transfer of anxiety. Increasing the professional's awareness of this transfer brought more effective involvement with the family. Brodey, a psychiatrist, noted that giving a social worker to a patient's family often resulted in conflict between the psychiatrist and the social worker. Brodey and Bowen both viewed this as upset moving from family members to professionals.

The theory takes on substance

Compared to findings in the earlier studies at Menninger, the closeness/distance cycles seen with the actual mothers and daughters were more distinct on the NIMH ward. The fluidity in the symbiotic relationship was unexpected. The original hypothesis described the oneness at a psychological level of "two people being one person, of twins who think alike, and of a husband who vomits when his wife is pregnant" (Bowen, 1957a, viii). But the observations at NIMH showed a deeper connection, "a state of two people living and acting and being for each other—of one being sick for the other to be well" (Bowen, 1957a, vii). Bowen's shift away from the psychological view now allowed him to reactivate his hunch from nine years earlier that emotional illness in humans is a similar process to what occurs in other living things. The level of interdependency went well beyond personality; this change in the theoretical understanding was an important contribution in a theory searching for evolutionary roots. The quarterly report was the first important nodal point in developing a natural systems theory (Bowen, 1978, xiv).

The new hypothesis

Bowen began plans for ways to bring in fathers to the project (Bowen, 1972a, 2). When fathers visited the family, interactional patterns included them. These fathers were long divorced from these mothers, suggesting that mothers had no significant support figures in their families (Bowen, n.d.c, 1).

> The question began to be asked about the family person who had this supportive relationship to the mother. The research mothers seemed to turn to other children, to their sisters or to a series of superficial community support figures, each used in a crisis and none of a continuing deep basis.
>
> *Bowen, n.d.c, 2*

The research observations of an integrated, interdependent unit brought a much stronger understanding that shaded into the biological. The original hypothesis underestimated the intensity between parent and offspring and the relief-seeking that came by engaging a third person. Psychotherapy had suggested that the symbiotic pair was a weak, undifferentiated, incomplete piece of a fuller group or larger emotional unit that gave relief and stability, not resolution. Neither family member showed an inner strength to separate into two autonomous people.

One can speculate that observations of the Genain family, who were living in the unit since January, added facts to be incorporated in the family unit theory, pushing the treatment plan to fit with the hypothesis. As Bowen was trying to shift his thinking from individual to family, he had to rework his treatment approach from two individuals to a single entity.

> My earliest working model of family therapy with schizophrenia was to leave the intense emotional dependence in the family where it had developed and to avoid the transference of the dependence to the therapist. I had specific techniques for this. I did not keep this model more than two or three months.
>
> *Bowen, n.d.b, 25*

This quote dates his transition to seeking a new treatment model to January 1955. Bowen was not reluctant to let go of a practice if it was not effective. Self-assessment and stepping back helped, as did having a mind open to such a possibility as far back as his hunch in the mid-1940s about the human. His search, his goal, was for an adequate theory of human behavior, saying

> my own ability to achieve a family unit viewpoint is in direct proportion to the working level of control over countertransference involvements. In previous individual psychotherapy, [I] had achieved what I had believed to be a workable level of countertransference control. When first exposed to the mother and patients, I found my countertransference control to be far from adequate for that situation. However, after a few weeks, there was the beginning ability to see the emotional struggle as a phenomenon. There was much less emotional participation in who won or who lost .... I was working hard to achieve objectivity but I believe it was the opportunity to see equal involvements by both mother and patient that accounted for the shift.
>
> *Bowen, 1958a*

In notes Bowen made during an interview with biographer Meryle Secrest about his discovery, he commented that "either the hypothesis was wrong or the application was wrong ... how to explain [the] behavior" (Bowen, 1972b, 3). A person's way of thinking filters what they perceive. He said that the emotional response of the therapist will define the therapy if a theory does not direct the clinician. Theory explains and guides. Bowen's interest was to use theory as a pathfinder "like Konrad Lorenz did with animals—part of man related to lower forms of life" (Bowen, 1972b, 3). To do this required a "broad framework that fits with other sciences." Bowen knew the exact moment when a new theory was born.

First time I saw this thing I had never seen before was exhilarating. After this my thinking time went to this … to get clear, it's a total preoccupation-trying to understand the phenomena–reading, finding models in nature that might fit …. Couldn't wait to tell others but they couldn't see it … [I was] amazed at numbers of people who were critical.

Bowen, 1972b, 4

Thomas Kuhn calls this the "… gestalt switch … that 'inundates' a previously obscure puzzle, enabling its components to be seen in a new way that for the first time permits its solution" (Kuhn, 1970, 122). Bowen's method for integrating discovery was that any conclusion must recognize human behavior as being based in evolution. The family unit fit with his original idea of emotional illness and fit with family groups in other species.

The Clinical Center was eager to host innovative research using a theoretical base. Bowen had two parallel tracks. The first, investigating how to move Freudian theory toward science (Kerr & Bowen, 1988, 351), the Clinical Center would welcome. The second track that a theory of the human had to be based in evolution and that psychotherapy must follow the premise might not have been as enthusiastically embraced.

The family unit discovery added a missing piece to Bowen's second track of thinking. It catalyzed Bowen's new unnamed theory. What he saw and heard on the ward and incorporated into the hypothesis convinced him that a more scientific approach was achievable. Evidence that supported his new theory was already in place, absent only this unifying component.

Once a more scientific understanding has traction, a long road of data gathering lies ahead to shift from possibility to confident knowing. Bowen started his move forward by designing more fieldwork to further generate naturalistic observations. He wanted to bring in intact families and find a treatment program to go with families. He employed the inductive theory building method described by theorists Glaser and Strauss in 1967 (as cited in Patton, 2002) "Generating a theory from data means that most hypotheses and concepts not only come from the data, but are systematically worked out in relation to the data during the course of the research" (Patton, 2002, 125).

To review, within five months of the project's start, the understanding moved from the specific details of the observations to the underlying processes explaining the patterns observed. These patterns formed the hypothesis that the family is a unit that functions as a single entity. Later, deep into writing up his original "six concepts," Bowen reflected on how the remnants of individual thinking before 1955 obscured his seeing the family as an interacting whole. Once he saw it, his focus went to finding these underlying processes in families with less severe problems to generalize the model (Bowen, 1964).

At Menninger, Bowen had observed and participated in symbiotic relationships, which he understood then to extend individual theory. By 1954, bringing the symbiotic pairs into a day-out, day-in living situation at NIMH

revealed a[n] … area of 'blindness.' The evidence of the larger family pattern
had to 'knock us down' before we could really 'see' it. The hypothesis … had
not even considered the way this interdependent two some would relate
to other family members who came to visit, or to members of the ward
staff …. Other family members were intimately involved in the emotional
system with the mother and patient, and the fathers were as deeply involved
as the mother. After it was possible to 'see' this family relationship system in
families with severe problems, it was then possible to see it in families with
lesser problems, and even 'normal' families. I think we often overlook the
value of schizophrenia in teaching us about the total human phenomenon.

Bowen, 1965, 7

For Bowen, the family unit idea meant multiple individuals deeply emotionally
connected, influencing their trajectory.

Origin of the family unit idea

There was an unexpected point made to me when Mrs. Betty Basamania, social
worker on Bowen's project from October 1955 to December 1958, came to the
Western Pennsylvania Family Center in Pittsburgh in 2002. She and Brodey did a
seminar on their experience on Bowen's project. Both worked closely with Bowen
on this project at NIMH. Mrs. Basamania wrote me after the event to correct something I had said during the day. She said the pivotal conception of the family as a
unit originated from her.

The research group was made up of Dr. Bowen, Dr. Dysinger, Mrs. Basamania
and Dr. Brodey when he joined the staff. The group met in Dr. Bowen's office
following each therapeutic session. Actions and observations were reviewed
and conceptualizations formulated from this data. It was about a year out
when including fathers in the study that the family came into consideration.
Since Mrs. Basamania was a social worker, she had studied and worked with
the family unit …. Due to this background, she suggested the concept of the
'family unit.' The psychiatrists were immediately interested because it was
appropriate to our ongoing observations.

Basamania, personal communication, January 24, 2003, 3

It is very possible that Mrs. Basamania's assertion was right. She joined the project in October 1955. Bowen was well aware of the critical assessments of others
on his project's continuance at NIMH. Bowen did not proselytize the "family as
a unit" understanding. He did state it in his March 1955 report, and he discussed
his thoughts around family in supervisory meetings. His actions indicate this
understanding as he began in March 1955 seeking families for the project and
searching for a suitable treatment method.

To have two researchers reach the same understanding using the same families, without one telling the other, is exciting. It supports the theoretical conceptualizations and lends credence to the research. The family's interdependence had biological and psychological ramifications, seen by another professional lent it support. "Science needs a community because we need others to correct and verify our perceptions, our hypotheses and interpretations of data" (Landaiche, personal communication, January 8, 2020). Bowen was well aware of timing in introducing new ideas. "Credible discoveries should be … able to be verified and validated by others—anyone, anywhere, anytime" (Grinnell, 2009, 66). Bowen himself documented this approach in a 1976 letter to E. O. Wilson "I left it to others to slowly discover the differences for themselves" (Bowen, 1976).

Deborah Weinstein, Ph.D., who used the archive materials for her book, *The Pathological Family: Postwar America and the Rise of Family Therapy*, wrote:

> About 'the family as a unit', I think that particular phrase was uniquely Bowen's, but the idea that the family functions as a whole, that it is a system and that it is the appropriate level of therapeutic intervention was more widespread. Bowen was not the only one to compare the family to an organism, for example. However, there were important differences in how early family therapists conceptualized how families functioned and how they attempted to intervene.
>
> *D. Weinstein, personal communication, February 21, 2003*

Bowen came to his understanding in his own way, and he was the only professional who based the idea in evolution. Others with whom he had relationships had their version of the idea.[2] There are correspondence exchanges between Bowen and each of these researchers. But the model of the family as a biological system, naturally formed under the influence of evolution, was solely his.

Other possibilities in the historical record

An interesting historical note is Nathan Ackerman's presentation, "Determinants of Disturbances of Mothering and Criteria for Treatment", at the American Orthopsychiatric Association's annual meeting on the morning of March 2, 1955. That date is cited online in many places as the original debate on family therapy and that Bowen was present at that event

> In 1955, Ackerman organized the first session on family diagnosis at a meeting of the American Orthopsychiatric Association (AOA). At that meeting, Jackson, Bowen, Wynne and Ackerman learned about each other's work and joined in a sense of common purpose.
>
> *Nichols, 2013, 21*

My interest in determining the accuracy of the statement is that 13 days after that meeting, Bowen used the phrase "family as a unit" in his NIMH report. Bowen answered the question of when he and Ackerman met in a letter in 1988.

> I heard about Ackerman when I went to Topeka in Jan 1946 .... In about 1955, he heard about me at NIMH and began to write. It was quite a relationship by mail. I had a definite theory (Topeka + Darwin + natural systems), and he was still fooling around with an empirical version of Freud .... I finally met him in person in about May 1957 at an APA Meeting in Chicago, a couple of months after the first family therapy nat[iona]l m[eetin]g.
>
> *Bowen, 1988, 1*

Whether Lyman Wynne, then co-investigator on Bowen's project, was at the AOA meeting is unknown. He is cited as being there in the quote above (Nichols, 2013, 21). Brodey was there and spoke in the afternoon session. He confirmed Bowen was not there.

> Bowen was not at this meeting. I did not know of him. By chance I heard of the Bowen unit from Robert Cohen ... my interest in family therapy made me call .... What a blessing!! He was interested in my family interest.
>
> *Brodey, personal communication, June 9, 2020*

Brodey became an investigator on Bowen's research project the next year (Bowen, 1956a). He came to the project attuned to the research efforts of Bowen and attracted by Bowen's discovery of the family as a unit and the search for a compatible treatment. Ackerman's ideas emerged from psychoanalysis; Bowen's from questioning assumptions in that theory. Both oriented to treatment for a family.

In March 1955, there was no treatment and no theory for families beyond what Bowen had tried already. But an undated review by Bowen in 1953 of a book, *Maternal Dependency and Schizophrenia: Mothers and daughters in a therapeutic group, a group-analytic study,* by Abrahams and Varon, showed Bowen's awareness of

> group psychotherapy in a group of mothers and their schizophrenic daughters. The other half of schizophrenia, the need of the mother for the helpless schizophrenic infant, is brought into sharp focus. This approach helps immeasurably in understanding schizophrenia. It may have value as a therapeutic method. Its greatest value may lie in understanding the normal mother-child relationship for where, but in investigation of grotesque pathology have we learned most about subtle nuances of the normal.
>
> *Bowen, n.d.d*

Staff suggested families could organize as a group and be part of the ward administration. The intent was for them to own what happened in the unit, to take responsibility for their own daily life.

Finding a treatment to fit with the new hypothesis

The idea of a group session on the unit came out of clinical necessity to bring organization to the disorder on the project and matching treatment to hypothesis (Bowen, n.d.a).

Staff's desire for instructions on dealing with certain incidents led to asking the patients what they wanted (Bowen, 1955b, 1).

The solution was a weekly talk with psychiatrists, social worker, occupational therapist, research assistant, nurses, aides and family members to discuss mutual problems. Bowen noted a "Patient grows from handling multiple concrete situations and not in abstract concepts" (Bowen, 1955c). Bowen emphasized self as the point of focus. It is action taken, "what is or what is not rather what I will do and not do–This concretizes an abstract situation" (Bowen, 1955c). Here were the earliest seeds of family psychotherapy.

Initially, the head nurse, or another designated staff member, would preside. There was much energy for this among all involved and the first order of business was establishing group policies of living (Bowen, 1955b, 1). This iteration was unsuccessful. There was no way to deal with the people who did not live up to these rules. The nurses became enforcers, creating a condition of feeling deprived for both staff and families.

Concurrent with starting a group meeting, nursing and other staff were working to develop a neutral language that avoided categorizing anyone, to help them be better observers and to strengthen management of their own reactivity. The Unit Reports showed substantial time in daily one-hour staff meetings went to reviewing and understanding the staff's emotional involvement in mother/daughter conflicts (Fisher, 1955a). The effort threatened continuance of the research as it used significant time with less than significant results. These daily one-hour meetings brought forth the open expression of feelings without solving the basic problems regarding "issues of ward management and nursing procedures" (Bowen, n.d.e).

The group reoriented by applying the expectations of families to the staff: observe self, learn about self, recognize when being helpful is not helpful and own your actions. It helps to imagine being a staff member on this ward. Staff had extremely unusual protocols to not be caretakers, but to observe what was going on between people, including self. The theoretical underpinnings of the project separated the staff from what they used to see as the trademark of their professions.

The therapeutic milieu group led by the head nurse, after experiencing inconsistent outcomes in March, made the psychiatrist the leader in April. Again, there was initial enthusiasm for this self-governing assemblage. Staff placed a suggestion box for ideas on topics of importance for discussion in the group meeting. Soon, patient complaints filled the box regarding staff and other family members.

Then the group selected its own rotating chairperson for a monthly tenure. That person needed inner strength to hold up against other group members' demands, and this did not materialize. Meetings stopped. Family members said, "We do not want to run the ward. We want to be patients. You (the staff) run it and we will not complain about any decision you make" (Bowen, 1955d, 2). Bowen's analysis was that the effort to aid the patients in forming a functioning group by having the personnel give them responsibility still left two groups: patients and the staff. This is reminiscent of Bowen's well-known saying, "You can't make a little bean grow by pulling on it" (Bowen, 1954).

Socializing the ideas

A benefit of exploring Bowen's archives at the National Library of Medicine is that it holds many drafts of his presentations. Some drafts read as if Bowen was free associating on how to speak from his now-different world to a conventional audience. Readers of the materials can get glimpses of Bowen's attempts to work through how to communicate this new way of thinking. In his drafts, he noted that his intent was to speak to a new point of view (Bowen, 1955e, 1). He indirectly referenced what he did in the March quarterly report. "If there is a bright new star in the realm of mental problems, it probably lies in the area of a new way of looking at things we have known for a long time" (Bowen, 1955e, 1). Another rewrite said "Some of our most promising concepts are new combinations of old information" (Bowen, 1955e, 2).

How would Bowen move forward with his essential conclusion about the family as a unit? In April, one month after he used the phrase in his quarterly report, he was one of three guest speakers at a "New Frontiers in Mental Health" conference in Bloomington, IL. The purpose of the meeting was to explore factors promoting quality mental health in communities. Kuhn recognized Bowen's position "… though the world does not change with a change of paradigm, the scientist afterward works in a different world" (Kuhn, 1970, 121). This meeting gave Bowen a space to assess how he operated in the conventional world with his new understandings. Could he convey the complexity of what comprises mental health as he now knew it, including differentiation of self and the family unit as an emotional system? Could he represent his position of being on the edge of complex questions, searching for a treatment for a family unit?

Adding another example of family to the ward

Bowen admitted one of his alcohol regression patients from Menninger to the 3-E ward in May. For this narrative, I refer to him as Mr. Etoh. Bowen had continued seeing him as an out-patient since leaving Menninger. The plan in having this person admitted was for his wife to live as his caretaker in the hospital (Bowen, 1955f, 3). It was a new version of the regression approach, adapted to family as the unit. Bowen's method was always to advance his knowledge through cross fertilization of different diagnoses and levels of impairment.

Here, he used different family configurations too. This month, Bowen had two projects, one with mothers and daughters with schizophrenia and one with a husband with alcoholism and his wife. With continued proximity between this couple, the wife developed physical symptoms that impaired her ability to "give" to her husband by being his caretaker. The symptoms kept her from remaining on the ward or even visiting. This was a repetition of the observations in the mother–daughters research. Increased physical closeness produced a physical symptom of distancing. Bowen understood distancing as one way a relationship system found equilibrium (Bowen, 1966, 154). Distancing kept the closeness in the relationship balanced. Said another way, rejection arose from heightened sensitivity to closeness. This is an area that Bowen identified for further research.

> Another whole area, still barely touched, and for which research data exists, is the area of somatic problems. We have referred to this as the way the 'soma of one family member reciprocates with the psyche of another.' There have been repeated examples in which, on this deeper level of family unity, this order of phenomenon can take place. None of this data has been worked up in terms of a research report.
>
> *Bowen, c. 1959*

The ongoing search for a treatment model

Staff tensions and high turnover among staff affected the milieu and interrupted Bowen's efforts to advance the integration of theory and treatment (Bowen, 1955g, 1). In his interim report in May, he requested the transfer of social work, nursing and occupational therapy services from general Clinical Center oversight to Adult Psychiatry supervision. He hoped to reduce tensions between these services, staff and the research project (Bowen, 1955h, 3). As for the group approach, it was appealing but in need of some revision. Bowen next made everyone on the project take part, with no closed-door discussions of patients by staff, and with patients as equal participants in the group. "Such an experience could not help but be maturing for both patients and staff" (Bowen, 1955f, 2).

By June, another try at a group meeting occurred. Individual therapy for each family member now made no sense unless paired with the understanding of the family as an emotional unit. "A research observer 'thinks' with the body of knowledge with which he is most familiar. One thinking model is as good as another for immediate usefulness" (Bowen, 1977). "Immediate usefulness" referred to a clinician having a platform for observing and assessing. Such a platform allowed for managing the clinician's anxiety.

Unless a model can account for exceptions, it loses its efficacy. There was no known family treatment alternative to individual therapy beyond what Bowen was doing at NIMH. Earlier attempts at group therapy had not worked out. In June, a series of incidents with a mother and various staff members led to thorny problems in the unit. Bowen described it thus: A mother could talk to a staff member, go back to her daughter with a distorted version of what the staff member said, and the daughter would act out the distortion (Bowen, 1957b). Observations, not interpretations based on the assumptions underlying a diagnosis, generated a rule to address this. Only the ward psychiatrist could discuss important issues with either mother or daughter. This was another effort that proved difficult to carry out.

As Bowen grappled with treatment issues, the decision to add fathers to the study advanced. By May, Bowen had extended his original hypothesis to include fathers and siblings. He reflected on this near the end of his life.

> There was not enough strength between the mother and daughter to make progress to a resolution of the way their lives were going. The inductive reasoning said to add the fathers, i.e. the family, would then become 'the patient.' This set the stage for the entire family to become 'the patient', when the details were worked out.
>
> *Bowen, 1987*

Hoping to give strength to the family and to put flesh on the bones of his new fieldwork, Bowen sought administration's approval this month to add fathers. It took six months to implement the change (Bowen, 1958b).

Between June and July, several refinements in ward policy took place. There was a resolution between Bowen and the staff that issues between a mother and daughter would wait until the family members and the staff observers were in the same room. It did not work. Then the rule became that no one spoke to either family member over these issues until mother, patient and all staff on duty were in the same place together (Bowen, Dysinger, Brodey, & Basamania, 1957). This was more effective. The transfer of tension went from staff to families too, not just the reverse. Bowen thought that the tension between him and the nurses around a workable group meeting was a source of the tension in the families (Fisher, 1955a).

These reiterations of a therapeutic milieu via patient and staff meetings served three purposes. One purpose was to understand how ward policies and procedures contributed to ward tensions. A second purpose was to open communication between all involved; it was important to have the family's reactions to these pending changes (Bowen, 1955j, 4). Bowen was genuinely interested in the family's response to the new direction of the investigation (new hypothesis, adding fathers). A meeting where anyone could speak on this made sense. Another purpose was searching for an effective treatment method for a family approach.

July 1955

While Family A was away on their summer pass, staff attended to the other two schizophrenia families, his alcoholic study family, the Genain quadruplets and the patients of other doctors filling the ward beds. Nurses paid attention to their own functioning. Bowen's families lived with two sets of ward operating procedures (those for Bowen and those for patients not in Bowen's project). Nurses wrote up interactions within families, with staff and between the families and complied with hospital requirements for the other ward patients. It was a high bar, but conse-quential to the investigation. The nursing staff answered to Central Nursing in the hospital, to the orders of the doctors and to themselves in executing their tasks of observing and recording.

For Bowen's project, all personnel, research and nursing staff were to be reading in the biological sciences while doing their jobs. Still, to the staff, the families were the best teachers in how a natural human system functions. That July, an example was when they recognized a pattern of resistance in staff toward Daughter C when the C family abdicated owning their relationship. When the staff could see their part in this problem, ward management changed so that the daughter's initiatives, not the staff's, took precedence (Bowen, 1955k, 2). This settled the disturbance in the family relationships. The family was no longer captive to what staff thought was best for them.

> In looking back over this one wonders why we had missed this area so long. [Daughter C] was quite tense on Thursday, July 28 … certainly this was a period when she had been separated from her mother for a long time. Our basic operating principle had been to do nothing which would interfere with the patient and the mother getting together any time it so worked out between them …. She is completely dependent upon the mother for spending money …. In other words we were not denying her contact with the mother, but she had to get money from the mother and had to keep enough money from the mother to be able to make these telephone calls. When she asked about a call on July 28, I told her it would not be possible because of our previous principle and the rule which I did not intend to violate. This was discussed at a staff conference on July 29 and we decided to change the whole thing to permit this patient and all other patients to use ward telephones for local outside calls. This came much closer to our old principle of permitting patients and mothers to communicate with one another, and certainly it took away from this mother the control she had over the patient's communications with her. Following this change on July 29, the patient began making many calls to the mother and bringing up issues with the mother that would otherwise have been left untouched. This increased communication resulted in the mother and two sons coming in for admission one week later.
>
> *Bowen, 1955k, 2*

This change showed the value of considering self as part of any problem. Somehow staff forgot the operating principle of keeping exchanges open between family members. Once staff saw their own over-functioning behavior operating in the month-long upset in the C family, they restored the principle of self choice. Nine months into the project, they needed recalibration for further progress, and the project's creators corrected what they had created. That brought significant dividends and reshaped the efforts of Bowen and staff to seek more objectivity.

A new selection criterion formed the families to be admitted next: The experiences with Family C, particularly the transferring of anxiety to another sibling, made Bowen cautious about having multiple sibling families until he could understand the process in small families (Bowen, 1956b).

An important connection

Though on vacation for half of July, Bowen's activities equaled a full work schedule. He visited psychiatrist Leslie Osborn[3] in Wisconsin who engaged him in an exercise of "role playing" (Bowen, 1955l, 1). Having given his consent to this before the meeting, he wrote, "I am in favor of any such device that can change an intellectual exercise into a live experience. So, count me in" (1). While visiting, their discussions ranged from Bowen's involvement in "therapeutic administration" (Osborn, 1955) to his interest in societal process. This intersected with Osborn's plans to "develop a program which emerges out of the social growing up of the people of this state based upon a long record of conscientious citizenship, mindfulness of the needs of those in trouble … and respect for integrity in government" (Osborn, 1955).

After this visit, Bowen invited Osborn to come and speak to others at NIMH (Bowen, 1955m). He agreed, visiting on October 5, 1955 (Bowen, 1955n). The relationship with Osborn developed into a long-term correspondence lasting until 1983, with visits when possible. The respect between them was significant enough that Osborn encouraged Bowen to come to work in Wisconsin upon leaving NIMH. Years later, Osborn wrote to Bowen, "I see you as one of the great phenomenologists of psychiatry, and medicine" (Osborn, 1983).

Ward dynamics

So far, this entire year reflected the messiness of the research. Patients on the ward kept changing, and the different protocols of admitting doctors challenged the varied duties of nurses. Yet, there was a wealth of data available from the observations. The severe impairments of Bowen's families kept nurses busy well beyond being observers. They were handling medical care, standard nursing offerings to calm patients, such as massages, warm baths, cold packs, staff-family interactions and uncertainty regarding the ethical issues of the doctor's expectations. Bowen was trying to track all of this, work out the magnitude of the recent developments in his own mind, and figure out how to talk about it to others hoping to add some fresh perspectives to his ideas. What seems so clear is that Bowen stayed with his central goal to plan a more factual theory and kept his footing on that trajectory. He wrote that he had thought out every possibility in the hypothesis so that there was a plan to address nonlinear outcomes and to not react personally. Theory building was the intent. Families and schizophrenia were the laboratory.

Another view representing his trajectory

Deborah Weinstein made note of how Bowen's clarity of expression changed. Her book, *The Pathological Family*, contrasted Bowen's statements in 1955 about adding fathers to what he wrote in 1959. In 1955, adding fathers was to give support to the mother while recognizing that the father-mother relationship was significant to the mother–child relationship. The underlying structure of the study was changing from mother–child symbiosis as "an isolated entity within itself" (Bowen, 1955h, 3) to a family study where the mother–child is "the axis in the development of the patient's incapacity" (3). By 1959, Bowen described that his scientific process "underscored the link between the production of knowledge and the production of a notion of the family itself ... analogous to an organism" (Weinstein, 2013, 124).

When Bowen wrote of the family as a unit for the first time in March 1955, saying it had great promise, it was an unexplored idea with opportunity. He could now revive his "background thinking from the 1940s about discrepant thinking models, and also the 'far out' hunches about the biological 'animal in man'" (Bowen, 1977). The family unit concept pulled together the other parts of the nascent theory. His explorations from this point forward confirmed his growing awareness of the family as a natural organism, a unit that interacts and influences its members deeply. This was the piece that met the demand for a biological rationale of a family based in evolution. Yet without confirming data, Bowen would open himself and his project to serious reactivity as "the idea about the relatedness between human and animal behavior was so taboo in the 1950s that the funders might have withdrawn their support if they knew the background ideas" (Bowen, 1977).

His writings show an immediate connection, an "Aha!" moment, when he first grasped the idea of the family unit. Years of thinking, observing, and hypothesizing diverged beyond a conventional understanding to manage or eliminate the troublesome behaviors. The family and staff responses on the ward were conceptualized to give a wider context for understanding human behavior within the family. That discovery restored his decade old hunch that the human was like other living things.

By 1956, Bowen had confidence that this new understanding could achieve his goal to "eventually elevate psychiatry to the status of the accepted sciences" (Kerr & Bowen, 1988, 345). To generalize his theory, Bowen was also seeing out-patient families during 1955, giving more validity to his observations. These were families with less severe problems, yet they had comparable patterns in functioning.

Communicating about thinking

Bowen again presented his April paper on July 7, 1955, at a conference on guidance and counseling, at the University of Wisconsin, Madison. It was polished and re-titled to "Practical Frontiers in Mental Health." The second presentation gave him an opportunity to hear his own thoughts in a public forum again. Bowen wanted more exposure to the response of interested others beyond his familiar peers at NIMH. This was a way of screening his ideas beyond the emotional field of the Clinical Center and triggering his creative process.

Some of the most important discoveries in the field of psychiatry and mental health will be found in new ways of looking at facts we already know .... This

> is a concept which says that some of our most formidable blocks to progress
> in understanding emotional problems rise ... in the minds of men and the
> inability of our society to face itself.
>
> *Bowen, 1955d, 1*

Bowen mentioned lag time for accepting new ideas was a product of a closed mind and of denial in society. He was still thinking about this thirty years later.

> A generation ago I wondered about 'lag time'[4] that was consistent with the
> Dark Ages and all the scientists since that time. That includes Darwin and
> Freud and Galileo who was expelled from the church some 400 years, for
> thinking wrong.
>
> *Bowen, 1984*

Already in 1955, Bowen knew that a more scientific theory would take years to be accepted. Here, Bowen challenged the audience to a new perspective with a broader base. "Is there a special category of physical health and another of mental health? If so, should we then also postulate other important areas–religious health–economic health–academic health–social health–marriage health–political health. How about heart health–stomach health–skin health–dental health–hair health–cosmetic health" (Bowen, 1955d, 3).

Bowen used Karl Menninger's 1950 paper to illustrate what he was proposing. Menninger had suggested "that the terms 'neurosis' and 'psychosis' be eliminated" (3); alternatively, "... a neurosis as a 'misadventure in living" is not a disease process (3). To illustrate, Bowen gave examples of a two-dimensional worldview of primitive man, that of length and breadth, a flat earth. The three-dimensional view added depth, a round world. Then Einstein offered a fourth dimension, time.

> When our capacity to think will permit them all to go together, then we
> will have developed to four dimensional reasoning. The world itself has not
> changed over the passing centuries, yet the way we see our world has changed
> a great deal (3).

Bowen was speaking about himself over the last nine years, about the experience of discovery, the difficulty of reasoning out ideas and having others hear and integrate the ideas with their own understanding. He offered other examples of shifting abstract processes in his thinking of the natural world, and of medicine and psychiatry. His research experience showed that blaming parents missed a much bigger point, and besides, it was not productive. He noted his attention to "emotional illness as part of the effort of the organism (a term used to denote both physical and emotional function) to restore itself" (9). Bowen then addressed this alternative view, a baseline for a family theory.

> Instead of symptoms being weakness and pathology, they constitute the
> strength that sustains life under adversity. It would go on the assumption

that there is a positive force within the human that works toward restoring a healthy state and that the job of the psychiatrist is to get on the side of and to aid this positive force (15).

Of note in this paper, Bowen used the term oneness, not symbiosis, to describe the long, involved separation between mother and child. He contrasted his family view regarding schizophrenia with the conventional understanding of schizophrenia as a basic impairment within the individual:

> The overall process of growing from a 'oneness' to a 'twoness' is long and complicated and … it is a phenomenon which resulted from a set of circumstances in which mother-father-and child and all others involved were doing everything in their capacity to prevent this happening and that the series of events comes about in spite of the best efforts of all concerned (19).

This presentation was Bowen's attempt to publicly describe the experience of thinking beyond a conventional viewpoint. Just as practitioners today must think about and apply systems in a conventional world, Bowen faced the same challenge in 1955. By talking about the process of scientific discovery with his peers, the presentations in April and July were an exercise in credibility. Did someone in the audience have similar observations, views and thoughts; or would they challenge what Bowen was saying or even inspire new thoughts about divergent thinking? Grinnell writes, "… talking about science can change an investigator's overall research approach … key comments and questions can play a critical role in determining how work on a particular research project evolves" (Grinnell, 2009, 70).

Sometime in 1955, Bowen also took part in a symposium on schizophrenia at NIMH, sponsored by George Raines, Chair of the Department of Psychiatry at Georgetown University. An associate of B. F. Skinner was the featured guest, and he presented on rat research where all the subject rats showed consistent findings except for one rat. When the researcher wanted to reject that finding, Skinner responded, "The rat does not lie."

> Strict adherence to this principle of looking to the 'rats' for guidance, and formulating theory from what the 'rats' said, and making decisions based on the rats and their theory led me farther than I ever hope to come when I started in 1954 .... When my own functioning is down, and I am as vulnerable as anyone else to making decisions based on clinical judgment and emotionality, I refrain from making any but the most temporary interim decisions until I can get my functioning back, and I can get back to the 'rats' and research and theory for guidance.
>
> *Bowen, n.d.f*

Bowen often repeated this refrain, "The rat does not lie" (Bowen, n.d.g). The principle of a theory that can account for exceptions served him well over the years.

Senior psychoanalytic consultant Lewis Hill[5] also has a place in Bowen's development of theory and concepts. Hill began at NIMH as an advisor to the institute at the exact point of Bowen's arrival there, in July 1954 (Bowen, 1957c). He was the consultant overseeing the exploration efforts at NIMH with the quadruplets since January 1955. In May 1955, Bowen submitted a specific, formal request for Hill to join his project (Bowen, 1955h), and Administration approved the petition in July. Interestingly, nowhere in that request is the family as a unit mentioned, but clearly Bowen had the idea and the consultant both in mind. Hill also served as Bowen's supervisor when he was completing certification as a classic psychoanalyst.

Bowen highly respected Hill and gave him credit for the idea of schizophrenia as a three-generation process (Hill, 1955). Hill influenced Bowen to venture "into the less conventional ways of going" (Bowen, 1963). Investigating processes that were occurring in families beyond the nuclear family was a requirement if in vivo experience shaped the family over time. Hill remained a consultant to Bowen's research until Hill's death in February 1958 (Bowen, 1960, 69).

Bowen accepted ideas from others beyond Hill or Skinner. Otto Fleischmann, from the Education Department at the Menninger Foundation, visited in April 1955, and his comments and observations inspired Bowen.

> One idea that has kept stewing around was the one that you mentioned about my 'marriage' to the head nurse and the need for this to be consciously recognized ....The ... need to find a new conceptual way to look at what we already know was something that fell on receptive ears.
>
> *Bowen, 1955o, 3*

Of note, this recognition of adjustments being needed in the professionals eventually led to the departure of this nurse in August 1955.

Societal trends

Psychiatric approaches underwent a renovation in the mid-1950s. Thorazine, the original antipsychotic medication, received approval from the Food and Drug Administration in 1954 (Pan, 2013). Soon thorazine was widely used in mental institutions. By 1955, the population of those institutions peaked at 558,000 (Pan, 2013). Miltown, the first tranquilizer for use by people experiencing anxiety, meaning ordinary folk never near a hospital, went on the market in 1955. Tranquilizers now accounted for a third of medications prescribed (Tone, 2012, xiv).

The environment in which hospitalized patients lived was also under scrutiny. Maxwell Jones, a psychiatrist from Belmont Hospital, Director of the Social Rehabilitation Unit at Brighton Road, Sutton, Surrey, England, published *The Therapeutic Community: A new treatment method in Psychiatry,* in England, in 1953. Psychiatrist Alfred Stanton and sociologist Morris Schwartz collaborated on the 1954 book *The Mental Hospital, A Study of Institutional Participation in Psychiatric Illness and Treatment,* produced in the United States. Both books proposed substantial change in the mindset of directors and ward staff, a rebalancing to a more open system. These authors advocated for a more democratic exchange integrating multiple voices and views.

> As differences of opinion between therapists, administrators and nurses were investigated … many times administrative, therapeutic or nursing procedures reflected personal needs of those involved rather than reality needs of the patients …. Multiple points of view embracing those of therapist, clinical and medical director, aides, nurses and supervisors supply a much broader base for intelligent clinical administration.
>
> *Stanton & Schwartz, 1954, vi*

Frieda Fromm Reichmann, Harry Stack Sullivan and William Menninger were the well-known consultants to this latter book. Stanton and Schwartz studied a female in-patient ward at Chestnut Lodge Hospital, where Robert Cohen, Ph.D., had been Clinical Director before coming to NIMH. A therapeutic milieu meant everyone there had responsibility for patient care, leveling the hierarchy and elevating the patient from the one-down position. Their study, done in the six years before publication of their book, coincided with Bowen's regression work. They based their efforts on the assumption of the individual having an impaired mother, an understanding that fits perfectly with a psychoanalytic approach. The findings paralleled two of Bowen's discoveries, the usefulness of a treatment contract in resolving what they describe as a "power operation" and creating an environment "to make up for the past deficit in the patient's rearing" (483).

> The authors' emphasis is on the social environment in a mental hospital. 'It forms', they say, 'the context of which the patient's illness is a part and is also an area of direct access to the psychiatric administrator'. The entire volume is indeed a study of the clinical administrative management of the patient's living 'during the other twenty-three hours', and throughout the book the importance of what goes on among the patients and between the patients and the medical and nursing personnel is emphasized as an important part of the evolution of the patient's disorder.
>
> *Overholser, 1955*

NIMH funded the Stanton-Schwartz study. It is unknown if Bowen read about this study (and it prompts speculation whether any consideration of this study played a part in bringing Bowen to NIMH). But broadening and extending understanding of what influenced human behavior and what environment worked more effectively using an institution model was in the air in the early 1950s. Bowen's unique insights moved him to an in vivo examination of the entanglement of the parent and offspring duo in a specific environment offering 24-hour observation. His research at NIMH was concurrent with the publication of Stanton and Schwartz's book, but Bowen made the quantum leap that the family might hold the answers.

Bowen's first try at a distinct family approach, having the mother and patient use therapy with the psychiatrist and social worker as a replacement for "the symbiotic struggle, did not seriously become used by either mother or patients as a symbiotic replacement relationship" (Bowen, 1990, 1). Mothers, even the one living in, acted more like visitors. It was such an advanced approach that the transition time for staff and families to grasp it required a year of practice. The period from 1954 to 1955 was a rehearsal.

Bowen knew that a visionary attitude had to be founded on practical application. The various trials of a patient-staff meeting since March were iterations of exploring this. For Bowen, the theoretical orientation was outside of the therapist, grounded in the known and observable facts and detached from the therapist's learned knowledge, training, experience, biases and intuitiveness. The treatment method had to match the theoretical direction.

The operating fact of a twosome was not enough people to contain upset. Bowen outlined his reasoning in bringing fathers and other family members into the project.

> Both (mother and daughter) can be victims but both cannot be separate individuals .... The decision to bring in F[ather]'s was the effort to find the family unit that could be autonomous and self-sustaining. They considered various family members including the mother's mother in the belief that mothers rel[ationship] to her mother must have had characteristics similar to the M[other] to the p[atien]t. The m[other]s in the project turned to various outside figures in the distance phases of her rel[ationship] with the p[atien]t. One mother had 3 other children at home and within hours after a distance separation from the pt. she would have another child involved in a symptomatic rel[ationship] .... Other displacement figures for M[other] were her sisters and her parents. It was decided that a mother and father who had maintained the family configuration separate from the families of either one or from other significant family figures would have the best chance of having the family problem contained within a small number and also capable of reassembling itself to adult autonomy.
>
> *Bowen, c. 1956, 17–18*

Bowen had observed family relationships beyond the mother and child. It was the dynamic occurrences associated with the visits of fathers and other family members to the project that confirmed his choice to add them. By the December 1955 annual report, he was ready to move in fathers or other close family figures (Bowen, 1955p, 3); presumably, this meant siblings though he had briefly considered going up the matrilineage (Bowen, 1979, 3).

A pivotal year

The first seven months of the project showed advances in the important realization of the family as a unit and in the developing multigenerational view of schizophrenia. Plans were forming to have intact families as research subjects. A shift from using the term symbiosis to the phrase emotional oneness showed Bowen's shift away from the jargon related to individual theory.

The year 1955 was an example of basic research advancing knowledge of symbiosis and schizophrenia. Bowen's personal "hunch" that emotional illness served as a connection between the human and lesser-developed animals was a possibility. Important understandings were in place related to individual emotional growth, differentiation of self and the emotional system. Those concepts were integral to a family unit. The thesis was changing, and how to adapt the ward and treatment to explore the hypothesis was a challenge. Kuhn writes that "all discoveries from which new sorts of phenomena emerge" have "characteristics that include: the previous awareness of anomaly, the gradual and simultaneous emergence of both observational and conceptual recognition, and the consequent change of paradigm categories and procedures often accompanied by resistance" (Kuhn, 1970, 62). And there was resistance in the administration and within Bowen.

The months from August to December 1955 would be a testimony to Bowen's fortitude and to the Clinical Center's trust in his integrity.

Notes

1 A 2006 review of this book by Monika Baege, Ph.D. is in *Family Systems Journal*, Vol. 7, No. 2, pp. 180–191.

2 Lyman Wynne, a colleague at NIMH, was given credit in an article in *The Washington Post* in 2007: "as early as 1947, Dr. Wynne came up with the idea of treating families as a unit, even when only one member was diagnosed with schizophrenia" (Sullivan, 2007). As referenced in Chapter XI, Nathan Ackerman used the term family as a unit and anticipated family psychotherapy in his 1950 article "Family Diagnosis: An Approach to the Pre-School Child" (Ackerman, 1982). Ackerman also compared the family to an organism. Ackerman's primary interest was in therapy, not theory development, and in adapting Freudian ideas to a broader base of human relationships. One of the oldest groups using a multidisciplinary team and advocating for change in psychiatry for considerations of family factors was Theodore Lidz and Stephen Fleck. (Dr. Theodore Lidz, a noted specialist on schizophrenia, dies, 2001). Don Jackson, in Palo Alto, CA initiated "the first family sessions ever reported" prompting further explorations in family therapy and was a later observer of the family as a system (Ray, 2004, 38). Cybernetics served as his frame of reference. It is easy to conceive that a cross fertilization took place over the years as each one heard the others or discussed their particular research interests with the other.

3 Leslie Osborn was the Director of the Division of Mental Hygiene for the State Department of Public Welfare, a Professor of Psychiatry and an acting director of the Wisconsin Psychiatric Institute in the University of Wisconsin Medical School.

4 Bowen uses this term to describe the period between a theory's successful introduction and its acceptance as a fact in the world of science (Bowen, 1988a, 351).

5 Lewis Hill was a preeminent psychoanalyst. He was Chief of Psychotherapy (Psychiatrist in Chief) at Sheppard and Enoch Pratt Hospital in Towson, Maryland; Assistant Clinical Professor of Psychiatry of Johns Hopkins; President of the American Psychoanalytic Association and the first director of the Washington-Baltimore Psychoanalytic Institute.

References*

*Unless otherwise specified, the works of Murray Bowen included in this chapter are from the Murray Bowen Papers. 1951–2004. Located in: Modern Manuscripts Collection, History of Medicine Division, National Library of Medicine, Bethesda, MD. The Accession, Box and Folder information differs for each one and that is included here.

Ackerman, N.W. (1982). In D. Bloch & R. Simon (Eds.), *The strength of family therapy, selected papers of Nathan W. Ackerman* (pp. 284–286). New York, NY: Brunner Mazel.

Baege, M. (2006). Patients have families, Henry B. Richardson, M.D. Family Systems Journal, 7(2), 180–192.

Bowen, M. (n.d.a). [Draft]. (Acc. 2007-073, Box 3, Folder Working papers from NIMH project).

Bowen, M. (n.d.b). [Outpatient family psychotherapy with the psychoses]. (Acc. 2007-073, Box 3, Folder M Bowen working papers re: NIMH project/psychoses).

Bowen, M. (n.d.c). [Chapter on the origin of the family movement and the rise of new methods of therapy in the 1940s]. (Acc. 2007-073, Box 3, Folder Working papers).

Bowen, M. (n.d.d). [Research on families with a schizophrenic offspring]. (Acc. 2007-073, Box 3, Folder Bowen-NIMH-working papers incl. proposal for project).

Bowen, M. (n.d.e). [Draft] (Acc. 2007-073, Box 3, Folder Working papers from NIMH project re schiz.).

Bowen, M. (n.d.f). [Draft of letter]. (Acc. 2007-012, Box 1, Folder Georgetown family center).

Bowen, M. (n.d.g). [Erosion of theory]. (Acc. 2007-012, Box 4, Folder Working papers).

Bowen, M. (1954). [Clinical notes, December 28, 1954]. (Acc. 2006-003, Box 6, Folder Formulation of family study project).

Bowen, M. (1955a). [Report on Research Activity of Laboratory of Adult Psychiatric Investigations, General Staff Meeting, NIMH, March 15, 1955]. (Acc. 2006-003, Box 4, Folder Report-NIMH general staff).

Bowen, M. (1955b).[Attachment to December 29, 1955 letter to Leslie Osborn. A dynamic conceptualization of the patient-staff group on a psychiatric clinical research ward]. (Acc. 2007-012, Box 4, Folder Misc.).

Bowen, M. (1955c). [Clinical Record, Unit Report, Bowen section meeting notes, March 29, 1955]. (Acc. 2006-033, Box 1, Folder March 1955).

Bowen, M. (1955d). [Presentation: Practical frontiers in mental health, July 7, 1955]. (Acc. 2006-003, Box 5, Folder Mtg. notes October 1954).

Bowen, M. (1955e). [Draft. Mental Health Institute,] April 12, 1955. (Acc. 2006-003, Box 7, Folder Mental Health Inst-Bloomington, IL).

Bowen, M. (1955f). [Project description: Investigation of the character-structure of the oral addictive character. May 26, 1955]. (Acc. 2006-073, Box 4, Folder NIMH–Annual report project description).

Bowen, M. (1955g). [Analysis of NIH program activities, project description sheet, "Influence of the early mother-child relationship in the development of schizophrenia", December 1955]. (Acc. 2006-003, Box 4, Folder 3-E project).

Bowen, M. (1955h). [Interim report on research project. "Influence of the early mother-child relationship in the later development of schizophrenia", May 23, 1955]. (Acc. 2006-003, Box 4, Folder NIMH–Annual report (project description).

Bowen, M. (1955i). [Letter to a Menninger colleague on May 31, 1955]. (Acc. 2006-003, Box 4, Folder Correspondence – personal).

Bowen, M. (1955j). [Letter to J. R. Rees on December 29, 1955 with attachment: Dynamic conceptualization of the patient-staff group on a psychiatric clinical research ward]. (Acc. 2007-073, Box 3, Folder Clinical notes on patient families in NIMH project efforts).

Bowen, M. (1955k), [Doctor's progress notes, August 22. 1955).] (Acc. 2006-003, Box 3, Folder (Daughter C. name redacted).

Bowen, M. (1955l). [Letter to Leslie Osborn, University of Wisconsin, May 19,1955]. (Acc. 2006-003, Box 4, Folder Correspondence – personal).

Bowen, M. (1955m). [Letter to Leslie Osborn on September 9. 1955]. (2006-003, Box 7, Folder Osborn, Leslie, Wisc. psychiatric inst.).

Bowen, M. (1955n). [Notes]. (Acc. 2004-043, Box 4, Folder Osborne, Leslie A., Wisc. Psychiatric Inst).

Bowen, M. (1955o). [Letter to Otto Fleischmann on June 3, 1955]. (Acc. 2006-003, Box 6, Folder NIMH Correspondence A through G).

Bowen, M. (1955p). [Analysis of NIH program activities: Project description sheet Influence of the early mother-child relationship in the development of schizophrenia]. (Acc. 2006-003, Box 4, Folder NIMH–annual report project description).

Bowen, M. (c.1956). [Clinical Record History-Part 1, March 21, 1956]. (Acc. 2006-003, Box 3 Folder Name redacted, Daughter A).

Bowen, M. (1956a). [Letter to George N. Raines on April 4, 1955]. (Acc. 2006-003, Box 6, Folder NIMH correspondence q through z).

Bowen, M. (1956b). [Draft. A psychological concept of schizophrenia]. (Acc. 2006-003, Box 8, Folder A psychological concept of schizophrenia).

Bowen, M. (1957a). [Presentation: Psychotherapeutic treatment of the family as a unit, February 1, 1957]. (2006-003, Box 2, Folder Drafts 1957–1958).

Bowen, M. (1957b). [Presentation Family participation in schizophrenia to psychiatric staff, Henry Phipps Psychiatric Clinic, Johns Hopkins University, Baltimore, Maryland, March 12, 1957]. (Acc. 2004-013, Box 4, Folder APA papers).

Bowen, M. (1957c). [Clinical investigations National Institute of Mental Health-Annual report]. (Acc. 2006-003, Box 4, Folder NIMH–Annual report • (project description).

Bowen, M. (1958a). [Draft: Family psychotherapy in families with a schizophrenic family member]. (Acc. 2006-003, Box 7, Folder Family research project notes and paper drafts).

Bowen, M. (1958b). [Draft: A family concept of schizophrenia]. (Acc. 2006-003, Box 4, Folder Chapter- a family concept of schizophrenia).

Bowen, M. (c. 1959). [Family Research Study—A Prospective]. (Acc. Acc. 2007-073, Box 3, Folder Bowen-NIMH-Working papers incl. proposal for project).

Bowen, M. (1960). A family concept of schizophrenia. Chapter 4 in *Family therapy in clinical practice*. New York, NY: Jason Aronson, 1978.

Bowen, M. (1963). [Letter to John D. Patton on December 23, 1955]. (Acc. 2007-014, Box 4, Folder Misc. correspondence).

Bowen, M. (1964). [Presentation: The Family as a Psychological Unit to GAP, Philadelphia, PA on April 20, 1964]. (Acc. 2003-044, Box 3, Folder GAP paper-April 1964).

Bowen, M. (1965). [Presentation: Theoretical and technical approach to family psychotherapy in office practice to South Florida Psychiatric Society, Miami, Florida on December 13, 1955]. (Acc. 2007-073, Box 3, Folder Working papers).

Bowen, M. (1966). The use of family theory in clinical practice. Chapter 9 in *Family therapy in clinical practice*. New York, NY: Jason Aronson, 1978.

Bowen, M. (1972a). [Letter on June 10, 1972]. (Acc. 2004-043, Box 5, Folder Family systems theory & death).

Bowen, M. (1972b). [Interview between Murray Bowen and Meryle Secrest on February 1, 1972], (Acc. 2007-073, Box 1, Folder Merrill Secrest].

Bowen, M. (1976). [Letter to Edward O. Wilson, Ph.D. on April 30, 1955]. (Acc. 2004-013, Box 3, Folder Family symposium, Oct. 21–22, 1976).

Bowen, M. (1977). [Letter to Professor Elof Axel Carlson on August 17, 1955]. (Acc. 2004-013, Box 3, Symposium 1977).

Bowen, M. (1978). *Family therapy in clinical practice*. New York, NY: Jason Aronson.

Bowen, M. (1979). [Letter on January 8, 1979 to Luciano L'Abate]. (Acc. 2007-012, Box 1, Folder Correspondence re: Publications).

Bowen, M. (1984). [Letter to a conference organizer, June 1984]. (Acc. 2003-044, Box 5, Folder Name redacted – Chicago, IL, June 14–16, 1984).

Bowen, M. (1987). [Draft]. (Acc. 2006-003, Box 8, Printed papers in odyssey, part 1 of 2).

Bowen, M. (1988). [Letter to psychiatric colleague in New York City on August 29, 1955]. Acc. 2004-013, B 4, Folder AFTA special 10th anniv. program).

Bowen, M. (1990). [Letter on April 22, 1990 to a psychologist). (Acc. 2003-044, Box 6, Folder Green Bay, Wisc.).

Bowen, M., Dysinger, R. H., Brodey, W., & Basamania, B. (1957). [Study and treatment of five hospitalized family groups each with a psychotic member]. (Acc. 2006-003, Box 6, Folder American Orthopsychiatric Ass'n.).

Brodey, W., & Hayden, M. (1957). [Intrateam reactions: Their relation to the conflicts of the family in treatment]. (Acc. 2006-003, Box 2, Folder REPRINTS-1952–1958).

Dr. Theodore Lidz, a noted specialist on schizophrenia, dies. (2001, March 2). Retrieved January 9, 2020, from Yale Bulletin and Calendar, Vol. 29, Number 21: http://archives.news.yale.edu/v29.n21/story14.html.

Fisher, T. (1955a). [Summary of monthly interviews with Mrs. A. from November 1954 through August 1955]. (Acc. 2006-003, Box 3 Folder Clinical record copies Daughter A., [Name redacted]).

Grinnell, F. (2009). *Everyday practice of science: Where intuition and passion meet objectivity and logic* (1st ed.). New York, NY: Oxford University Press.

Hill, L. (1955). *Psychotherapeutic intervention in schizophrenia*. Chicago, IL: University of Chicago Press.

Kuhn, T. (1970). *The Structure of Scientific Revolutions* (2nd ed.). Chicago, IL: The University of Chicago Press.

Kerr, M., & Bowen, M. (1988). *Family evaluation: An approach based on Bowen theory*. New York, NY: W.W. Norton.

Nichols, M. P. (2013). *The essentials of family therapy* (6th ed.). Boston, MA: Pearson/Allyn and Bacon.

No Author. (1955). [Draft, March 9, 1955]. (Acc. 2006-003, Box 7, Folder Misc. papers).

Osborn, L. (1955). [Letter from Leslie Osborn to Murray Bowen on September 16, 1955]. (Acc. 2006-003, Box 7, Folder Osborn, Leslie A., Wisc. Psychiatric Inst.).

Osborn, L. (1983). [Letter to Murray Bowen on July 28, 1983]. (Acc. 2006-003, Box 7, Folder Osborn, Leslie A., Wisc. Psychiatric Inst.).

Overholser. (1955). *Review of The Mental Hospital: By Alfred H. Stanton, M.D. and Morris S. Schwartz, Ph.D. New York: Basic Books, Inc., 1954*. Retrieved July 9, 2021, from The Psychoanalytic Quarterly: www.pep-web.org/document.php?id=paq.024.0591a

Pan, D. (2013, April 29). TiMELINE: De-institutionalization and its consequences. *Mother Jones*. Retrieved November 19, 2019, from www.motherjones.com/politics/2013/04/timeline-mental-health-america.

Patton, M. Q. (2002). *Qualitative research & evaluation methods* (3rd ed.). Thousand Oaks: Sage Publications.

Ray, W. (2004). Interaction focused therapy the Don Jackson legacy. *Journal of Brief Strategic and Systemic Therapy European Review, 1*, 36–45.

Roy, R. (1988). 'Psychosomatic families': Henry B. Richardson. *Journal of Family Therapy, 10*(91), 9–16. Retrieved November 28, 2020, from On-line Library Wiley: https://online library.wiley.com/doi/pdf/10.1046/j.1988.00296.x.

Stanton, A. H., & Schwartz, M. S. (1954). *The Mental Hospital: A study of institutional participation in psychiatric illness and treatment.* New York, NY: Basic Books.

Sullivan, P. (2007, January 21). Family Therapy Pioneers, A directory. *The Washington Post.* Retrieved June 6, 2020, from www.aamft.org/members/familytherapyresources/articles/08_FTM_05_23_60.pdf.

Tone, A. (2012). *The age of anxiety: A history of America's turbulent affair with tranquilizers.* New York, NY: Basic Books.

Weinstein, D. F. (2013). *The pathological family: Postwar America and the rise of family therapy.* Ithaca: Cornell University Press.

14

TOWARD THE NEW HYPOTHESIS AND TREATMENT

August to December 1955

Traditionally, August is a time when psychiatrists vacation. In contrast, on Bowen's ward, it was an exceptionally productive month. Bowen made two important decisions. One was to try another group treatment effort; the other was to take action to find families with fathers to add to the study (Bowen, 1956a).

On the ward in August

Family interactions on the ward were intense, though Family A was on a summer holiday. Unaware as yet, Daughter C is pregnant. Her next younger brother, the one their mother wanted admitted, spent the last 18 days of August visiting his sister daily and attending activities with her. Bowen thought of having the entire family live in then, but a limited visit was preferable (Bowen, 1955a, 2). In an interview with Bowen, done at the request of the young man's school, the son/brother gave his understanding of his situation.

> Mothers have to get their understanding from women …. My mother is trying to get this from me and it is killing me. I cannot stand it. There has to be some kind of way for her to get the understanding without making me feel that I am responsible for her unhappiness.
>
> *Bowen, 1956b, 7*

With her son at the hospital every day, Mother C did not visit her daughter. Her unavailability combined with unpredictable events such as the announcement the head nurse was leaving and the special nurse assigned to Daughter C was being replaced without advance notice. This young woman again put a bobby pin in her arm. She functioned as the shock absorber in her family and on the unit. While a predictable environment would minimize the expression of symptoms,

DOI: 10.4324/9781003027287-15

there was much unpredictability on Ward 3-E. Her pregnancy occurred under those conditions.

Third iteration of a group approach

On August 8, Bowen reviewed with staff his recent visit to Wisconsin and his contact with Leslie Osborn, a herald of the family psychotherapy movement.

After that meeting, Bowen found he had renewed energy to try another family-staff joint meeting. Bowen and the staff had spent as much as an hour a day in staff meetings to understand the problems on the ward. Though staff openly expressed their feelings, the problems continued. Bowen's interest was in how everybody in the environment, not just the families, contributed to the milieu. Many hours had gone into efforts to reduce intra-staff tensions and their faulty perception that ward problems belonged to the research families. As evidence of systems thinking, he understood the staff perceptions as a sensitive indicator of unaddressed staff tensions. Then families picked up these tensions, and in a feedback loop, felt blamed and reacted. Much of the intense feeling between family members and staff centered on minor points around ward management and nursing procedures (Bowen, c. 1960).

Bowen began a very low key, open approach to making another effort at a group meeting. He placed a notice on the ward bulletin board saying that on August 9, 1955, he would hold the first of a weekly staff meeting in the lounge (Bowen, 1958). Any interested family member could be there, but there was no requirement to attend. Because it was in a communal space, not behind closed doors, it attracted people to attend. This weekly gathering was an immediate success.

It was the third effort to find an effective treatment method, and he based it on his inductive reasoning of "seeing schizophrenia as a manifestation of a distraught family" (Bowen, 1955b, 3). In reflections, 34 years later, Bowen wrote

> The theory was fairly complete at NIMH, after July 1954. The talk was then as much technical as it was theoretical. Two research devices were employed: 1. to integrate family findings with the concept, and 2. integrate the theory and the therapy.
>
> *Bowen, 1988a*

At this early stage, the effort to relate to and treat the family unit generated resistances in the staff. The difficulty of transitioning from thinking of the family as a collection of individuals to thinking of the family as a unit turned their heads around. It was routine for staff to think in terms of the individual. So when Bowen said these early attempts were "family group therapy," he was distinguishing this family group from the later refinement to treatment of a family unit the next year. Reflecting on this shift, he wrote:

> According to our hypothesis, a psychotherapy based on individual theory and directed to a group of individuals in the same family would be "family

group psychotherapy," which is quite different from the method "family psychotherapy". The term family psychotherapy is to refer to psychotherapy directed at the hypothesized emotional oneness within the family.

Bowen, 1975, 75

Bowen noted that even within himself, there was resistance.

I had the strongest misgivings about the idea of 'family psychotherapy. At that time, I still believed that the only way to emotional maturity was the careful analysis of the transference relationship between the patient and the therapist.

Bowen, c. 1964

His doubts indicate the experience of one treading into unfamiliar territory. An individual therapy for mother and daughter, while quite effective, did not bring the expected change between them. Ordinary routine exchanges between the twosomes easily transferred into conflict between a family member and staff member. Then it transferred to tension between two staff members, then administration got involved (Bowen, Dysinger, Brodey, & Basamania, 1957).

Observation of the entire group together provided a picture of the way staff reenacted a family problem. "For the first time the family-staff tensions were on a workable level and the staff did not have to resort to rules and force to control the situation" (Bowen, 1958). This effort was successful in generating one group for patients and staff (Bowen, 1955c, 2).

This meeting allowed for a way to track, by direct observation, the rhythm of anxiety moving in the family, between mothers, daughters and staff, and within the staff. When conflicts from a family moved to staff, the family calmed and intra-staff tensions developed. The meeting gave the nurses and aides a mechanism to avoid intense entanglements with family members or with each other. It gave mothers and daughters a forum to address their problems. The meeting was a learning opportunity for a mother or daughter, as each became an observer of other families working on their own dilemmas. Families could directly observe the staff address relationship boundaries with each other and with other families. It was possible to see, in vivo, when upset transferred from family members to nurses in the meeting and to discuss it at once. Staff accepted that they were role models for families on how to work through differences and tried to do so. Families could observe a model to use in their own relationship.

This meeting met the criteria of three operating principles of the project: openness, the expectation that staff work on self, and treatment must match the hypothesis. Open communication went back to Bowen's Menninger's years. There, Bowen had taken notice of published writings of patient's personal accounts of psychotic experiences and how openness benefitted them, and he encouraged others in this direction (Bowen, n.d.a). These daily sessions were a serious effort to erase any remnants of a closed communication system. It was a model for families. Bowen believed that a family with access to important information, available to all members, was healthier than a family that withheld information and had secrets (Bowen, n.d.a). Openness meant that family members had unrestricted access to all records kept by the staff and permission to attend administrative meetings related to the clinical or research operation. It also meant if Bowen was presenting material about the project outside the Clinical Center, he went over drafts of his presentation at these meetings.

Bowen was directly exploring treatment methods to adapt for studying the family as a unit, as these records attest. "I developed my own method of family therapy during the Summer of 1955 (I had never heard of it before)" (Bowen, 1985). This was the trial run for "my already worked out plan for family therapy … operationalized in [December] 1955 when my first full family was admitted" (Bowen, 1985). The August ward meeting allowed discussion that would usually be out of a patient's hearing. Staff comments on a family member's dynamics, staff frictions and any third person conversation about patient problems were now open to all to hear (Bowen, 1955c, 5).

This ward meeting, with family members present, served as a replacement for all other meetings. Nurses would not address important issues with parent or daughter until mother, daughter and all staff on duty were in the same room. They brought any confidential conversation between staff and a family member outside of the meeting into the meeting, a Maxwell Jones idea (Bowen et al., 1957). If staff attended other meetings, they gave reports in the ward meeting. Families could take up personal or ward issues with the staff. This was the initial significant shift toward family psychotherapy. Practice taking up issues with staff soon transferred to doing that between family members.

The group meeting took hold quickly. A turning point occurred when nurses disclosed their professional struggles. A nurse noted that they could not give as much attention as the families seemed to need because of the heavy responsibility of the nursing duties. Family members related their difficulty in receiving as much nursing attention as they wanted. Soon, a daughter moved to a solution and asked how she might ease the burden on the nurses (Bowen, 1955c, 3). This led to a plan to have willing family members meet with nursing staff in selecting activities. The privilege of participating with nurses in the nursing station had an effect of feeling part of running the ward, feeling special. "This group who felt able to be 'mature for a day' really received more support in the meeting with the nurse than in any other situation" (3). Within weeks, nursing staff took notice of the change. "A nurse remarked that she had never seen patients mature so rapidly in such a short time. A second nurse said she had not seen nurses mature as much in such a short time" (5). This is a good example of a person's ability to function upward depending on a relationship and the environment supporting it.

Bowen noted it was the family members who adapted quicker to the meeting than the nurses did. This makes sense. The families stepped up while the nurses were stretching their limits. It was equalizing. This meeting emphasized helping the ward function better rather than helping the hospitalized individuals function better. It freed participants to speak about their concerns without restriction. The group was efficient for staff, as all other encounters, including ward rounds, ended with the establishment of this group meeting. "In a short period of operation, this kind of scheme seems to have a wide potential for making it possible for the most mature function of people to become operative" (Bowen, 1955c, 5).

What made this meeting work when the two previous attempts at staff/patient meetings (in April and June) had not? My supposition is it was the intervening time for Bowen to think further and the meeting with Osborn in July refreshed Bowen's thinking and clarified what he wanted and how to do it, including not forcing families to attend. This fresh approach partly avoided the complications of a staff member having an individual relationship with any one family member. A family member's increased involvement with staff also avoided taking on issues with their own family member. It takes time to adapt to new situations. By the August attempt, everyone on the ward had reflected on what was being tried, understood the rationale and was more accepting. Bowen integrated theoretical planning and treatment success into this iteration, making the ward more livable for both staff and families.

Staff now had a place where one staff member could address another staff member about over-involvement with a family. On hearing this, the families directed attention to discussions about boundaries. They could learn from the staff's discussion and then integrate this new knowledge for their own behavior. The meeting was to raise the awareness in each family of their particular problem-causing actions and reactions and to heighten observational abilities in nurses. While families differed, the underlying operating patterns were similar—a spillover of anxiety to others, the inability for a parent to say no, cycles of closeness and distance, symptom response to increased physical closeness and their functional interdependence.

The meeting gave a very broad view of where problems were occurring in all the relationships on the project. A lack of concealment and working on self were early precursors to an understanding of a better functioning society. The notion of staff paying attention to self was a point of differentiation in an emotional system. Change in one member influenced change in another. The family-staff meeting was a much better treatment match for the new hypothesis. But this meeting went beyond trying out a therapy program to support the new hypothesis. Bowen also intended it to reduce the chaos in the project and to calm the worries of administration about this project. It was a trial period. Bowen had not yet committed completely to this method as his choice of treatment. That took five more months, when this daily meeting led to family psychotherapy, the treatment choice in January 1956 for the initial intact family.

August presentation at Chestnut Lodge

Bowen was in the vanguard of involving families and staff in the treatment process.

Maxwell Jones and Robert Rapoport, a social anthropologist interested in family, published "Administrative and Social Psychiatry" in *The Lancet*, in August

1955. In this article, they reported on an investigation at Chestnut Lodge, the psychiatric hospital eight miles from National Institute of Mental Health (NIMH). The article compared practices there with practices in England. The authors identified anxiety transfer from staff to patients seen in treatment facilities. They cited the contribution of transparent communication between staff and patients to improved patient outcomes. The authors described a set of principles about relationships between "administration and treatment staff, psychiatric and nursing staff, staff and patients" from Chestnut Lodge (Jones & Rapoport, 1955, 2). Jones, who instituted social psychiatry at Belmont Hospital in England, also wrote that in England they admitted relatives "to the hospital for treatment along with the 'case'" (Jones & Rapoport, 1955, 5).

Bowen received an unexpected invitation to present his research project at Chestnut Lodge. On August 19, four days after the invitation, he and two of the research staff, nurse Marion Bowe and social worker Thais Fisher, presented on Family A in "Clinical Presentation to Clinical Investigations Conference." Each staff member represented their own perspective.

This was the first substantive presentation outside the NIMH Clinical Center on Bowen's research. The invitation coincided with the publication of the Jones and Rapoport article, which extensively referenced the Stanton and Schwartz study at Chestnut Lodge (discussed in Chapter 13). What prompted the invitation is unknown. It is a fact that Robert Cohen, who now oversaw Bowen's project, had been Clinical Director there. Daughter A had been a patient there. Finally, Bowen's was an all-female ward, with the exception of the sporadic stays of the Genain family, as was the ward in the Stanton/Schwartz study.

Bowen's presentation identified the project's operating principles and that they were based on staff observations of the way family members functioned during the months of hospitalization. In describing how the project was operated, Bowen and the research staff introduced the hypothesis of the family unit: that change in the mother–daughter relationship would alter the symptoms of schizophrenia. Bowen reframed the theoretical understanding from being a deficit in the offspring to being a family issue involving parents and child. The parents may have a physical relationship and appear to be functioning well to outsiders, yet a "real emotional union" is lacking. The child offers to the mother the emotional connection missing with the husband. Mother holds on to the child while wanting the child to grow separate. The entanglement persists, keeping both deeply oriented to each other. Bowen described the father as turning to other relationships such as in business, often doing very well with a high emotional investment there and not in the marriage.

> Our hypothesis would further say that the biological goal of both mother and child is that they both reach states of emotional maturity and that the arrested developmental process will again begin to take place if the emotional deprivation which made the arrest necessary, is satisfactorily supplied from other sources
>
> *Bowen, 1955d*

This was the challenge for the staff to meet the deprivations with support, not advice. The helplessness manifested in both family members. Their calls for the psychiatrist, hospital staff, even the hospital administrators to set rules and to advise how to handle the conflict between mother and daughter were symptoms of the "arrested developmental process." This behavior permitted mother or daughter to avoid the inner uncomfortableness around taking a stand and possibly jeopardizing their relationship.

The mothers had an extreme sensitivity to being perceived as ill. Bowen described how patients interpreted a chance remark from staff as a directive. He discussed the complexities of having a mother at the hospital. She was not a patient, so she required a different designation. "… The technique has been to regard mother as somewhere in the middle between a patient and staff … mother assumes some duties for actual care of the patient that would ordinarily be done by nurses …" (Bowen, 1955d).

Bowen used the phrase "intense oneness" or "emotional oneness or fusion" as a more exact depiction of the relationship process than was the term symbiosis. But he was cautious, using phrases such as "It is not possible to have any conclusions …" and "… tentative impressions …" and "the pattern of response … suggests …" and "there may be a therapeutic and conceptual advantage" (Bowen, 1955d). He did not spell out or name his idea of the family unit. Bowen described, without using jargon, the family psychotherapy method then under construction as an alternative way of thinking of human behavior and psychiatric treatment.

This was Bowen's opportunity to shine a light on his thinking as it applied to relationships and to convey the human as a biological being. His presentation related an understanding well beyond the field of schizophrenia. Bowen said this later became an issue that contributed to his termination from NIMH. The Institute studied schizophrenia and Bowen was generating a new way to explain human behavior.

Redesigning the research

The two visitor mothers in the project families did not take over actual caretaking duties to the full extent possible. Family A reached a plateau of functioning and Mrs. C shifted her over attention to a son at home, leaving her daughter on her own. Mrs. B, the one live in mother, had impairment so great she rarely acted as the mother. And when the daughter filled this position, she could not sustain it. These observations pointed to the need for a support figure for the mother and the best choice of a support figure for the mother was a husband (Bowen, c. 1957, 13).

Clinical necessity again pushed the move toward an integrated theory and therapy. Having a husband on the unit should reduce the need for the staff to be the third person. It was a step toward exploring the hypothesis that schizophrenia was a manifestation of a disturbed family. Two major theoretical assumptions were in the background: (1) that psychosis (including schizophrenia) is a dynamic intrafamily effect rather than a psychopathological entity within one person;

(2) that if the central figures in the family constellation are moved as a group into the therapeutic environment, and if the therapeutic environment can relate successfully with the family problem, the family disequilibrium can reverse itself (Bowen, 1956a). There was no data on a family method. Bowen was in foreign territory. But the data from the research project led Bowen to make an inductive leap and try this method of using intact families. Indicative of Bowen's creativity and intelligence, as well as having a true research mind, he was concurrently having success with two clinical control cases in his ongoing work with psychoanalysis (Bowen, 1959a, 2).

Having designed a viable family-staff meeting, Bowen's hunt for a treatment method was settled for now. What remained was to bring in intact families. During their monthly meeting, Lewis Hill mentioned the possibility (Bowen, 1956b) of a father–mother-impaired offspring-normal sibling family. On August 30, Bowen received the referring physician's referral, which described serious impairment in father and daughter (Referring physician, 1955). The next day, he phoned the institution to start the transfer process.

He was treading into unknown territory and knew the challenges he might face. Bowen followed the facts of the theory, not the traditions of clinical practice. All three mother–daughter pairs had shown "it is much easier to maintain the already existing symbiotic 'oneness' with the child than to maintain the relationship with the father" (Bowen, n.d.b, 25). Would that hold up in intact families? There was no precedent for this. Credit must go to Bowen's cautiousness.

> New theoretical thinking suggested a change to a family orientation, but practical issues favored a continuation of the individual orientation. Family theory was poorly developed and family psychotherapy seemed incomprehensible. One does not take lightly the clinical responsibility for such an operation. Individual theory, individual psychotherapy were within known and accepted areas of clinical practice.
>
> *Bowen, c. 1957, 5*

In individual theory, the model of relationship systems was transference between the psychiatrist and patient, which family contact could jeopardize. The dictate was "don't contaminate the transference relationship system" (Bowen, 1986, 2). If Bowen could adapt group therapy and apply it to a family, it could be a viable treatment. However, a researcher has to have validity in the research environment by speaking in terms that are understandable by his peers and those who oversee his operation.

Today, Bowen's vision and innovations can seem commonplace, but his policies were innovative: an unlocked ward, patients who read, and even wrote in, their own records, and staff who worked on their own maturity. His search for a more effective milieu was as groundbreaking as the trend then developing to improve institutional care and the lives of emotionally distressed families. Mental health laws were now releasing patients to return home. Society was addressing an expanded view of the institutional environment in psychiatric facilities and considering hospital staff and patients as a community.

Bowen held steady, implementing the new proposed hypothesis and continuing the group treatment approach to manage family–staff relationships. He also maintained his original plan of individual contacts between psychiatrist and patient and social worker and mother with the families already on the project.

Bowen had worked on how to keep family issues within the original relationships in his out-patient work back at Menninger. There, the parent and patient each had individual therapy contacts. At NIMH, he tried a hybrid approach. He increased physical proximity between mother and daughter by having mothers live on the ward or be present as much as possible. The enriched environment and having staff stay out of family processes created a cocoon that would leave the transference in the original relationship. It was not enough. Family upset still spilled over to others in the environment, a newly recognized fact that advanced Bowen's knowledge of how a family functions. When this happened, without the awareness of family process, staff considered one or both family members incorrigible.

Two elements were being explored: how to research a family as a single entity and how to provide the environmental conditions that supported this (Bowen, 1956c). Staff needed a new mindset for a workable treatment approach for both staff and family.

September 1955

The ward census for September included the mother and daughter in Family B, Daughter C, the Genain quads, their parents, who were admitted for eight days, and three of Dr. Scher's patients (NIMH Clinical Record, Unit Report, 1955b). Daily family–staff meetings continued this month and the Genain family attended. This family lived on the ward and took part in the unit meetings. Staff saw their family interactions, sister to sister, daughter to parent, and as they engaged with Bowen's research families. Most likely, Bowen watched them too.

On the first day of this month, Bowen followed through with the family referral received August 30. He contacted the Medical Director at the nearby psychiatric hospital and expressed his interest in the "schizophrenic daughter, manic depressive father, mother triangle" suggested for his program (Bowen, 1955f). The hospital described the seriousness of the father's depression, relaying it was not safe to keep him in a closed ward setting even after he received electroconvulsive therapy. At another hospital out of state, the daughter had been in a regressed, assaultive panic for four years and had massive shock and drug therapies (Referring physician, 1955).

The mother and younger daughter, who were at home, agreed to the entire family being hospitalized in December. The treatment plan would be up and running, waiting to be applied for the first time to an entire family. Staff observations had already established that the relations between mothers and daughters changed after a visit from father. Having the father live in could extend this understanding. He would provide someone other than staff for a mother or an offspring to turn to while the now-operating group meeting intended to keep the unresolved attachments in the whole family (Bowen, 1956b). Also of interest was the question of how easily would a mother's focus shift to a sibling rather than to the schizophrenic family member, as happened in the C family, when the father is present?

Changes in ward structure, departures and replacements

Personnel changes addressed issues that were unresolved with training and discussion (Bowen, 1955e, 3). The head nurse left the project in August for a different one within the Clinical Center (NIMH Clinical Record, Unit Report, 1955a). This had been brewing since April. It is unknown if this was connected to issues around the family-staff sessions, the disorder on the unit, the staff-administrative tensions, personal issues with Bowen or the head nurse's own personal issues. People self-select their tolerance level. When she left, no person or date was set for her replacement.

September proved to be a month of departures. Cholden, the friend who had influenced Bowen's decision to go to NIMH and who had brainstormed with him on the design of the project, left NIMH (Bowen, 1955g). Cholden moved to the University of California in Los Angeles (Gilliland, 1957). Tragically, he died in a car accident seven months later.

Cholden's departure was the first in a series of partings initiated this month. Mrs. Fisher announced her departure on maternity leave. When she left on October 28, Mrs. Betty Basamania replaced her. It was a timely transfer and there was no gap in social services for the families. In contrast, it took over a month to replace the head nurse. Mrs. Fisher's discussions of her leaving with each family showed a disciplined and determined effort to relate to the adult side of each family member. By focusing on the reality that this was a maternity leave, she gave her time and attention without withdrawing from the mothers. There were affectionate responses toward her from family members and a few abusive ones, too. The families faced another layer of managing self now, knowing they would have to continue with her replacement.

The administration's uncertainty about selected research projects, including Bowen's, played out in this last quarter of 1955. As the transition to operationalizing the family theory project approached, Bowen recognized the need to reserve his energy for this project. The duties of Ward Administrator were pressing into his clinical research time (Bowen, 1956b). The administration assigned Jordan Scher to replace Bowen in this position. How this selection was made is unknown. Scher took over the 3-E ward's administrative management duty on September 8, 1955 (NIMH Clinical Record, Unit Report, 1955b), the day after Cholden left.

The administration's stated reason for this change was to give Scher experience with the Genain family. The daughters were again on Bowen's ward and the parents were visiting this month, staying from September 9 to the 17 on ward 3-E. There may have been more to the choice of Scher than having him oversee the Genain family. Scher understood his selection as Ward Administrator was to bring order to the unit. Scher had a negative view of Bowen's approach and seemed to be an explicit threat to the milieu Bowen was working to maintain.

The direction of Bowen's project was poorly understood by others. High staff turnover plagued Bowen's project the first year. The project's design was in flux, as new observations guided changes to the hypothesis and the treatment method. Bowen was exploring a scientific theory while still relating to colleagues in their Freudian world. The 24-hour-a-day nursing notes confirmed the difficulties and commotion on Bowen's research ward. Despite an interest in fostering ground-breaking ideas, the administration ultimately answered to its funders, Congress.

Scher also brought patients from his research study to the ward. Brodey, a coinvestigator on Bowen's project, gave a neutral assessment, saying that Scher's project was an interesting one alongside Bowen's. It was different in substance and form—at an opposite end on a continuum (Brodey, personal communication, January 17, 2002). Each project on the ward distinguished itself. But each brought problems to the mindset then operating within the Clinical Center. A specific research direction for the Genain family remained elusive.

Two days after Scher's selection, Bowen wrote, "A previous administrative action abolished the project milieu on September 10, 1955" (Bowen, 1955e). He wryly noted, "It is a factual observation that some administrative orders transmitted to us have been transmitted in almost identical tones and words with which our most disturbed mother transmits her corrections to her daughter " (Bowen, 1955d).

In 1957, Cohen offered the position of the chief investigator on the research to Brodey. He declined and said later that the "National Institute of Mental Health wanted the families out" (Brodey personal communication, January 17, 2002). Brodey said that in offering him the job, the administration showed interest in supporting and continuing the project. Cohen assigned Brodey to get Bowen to write because the administration needed him to publish more. They were responsible to the funders who invested all this money in the project. Complaints from other staff, other research projects and the community regarding the families on Bowen's project were coming to their attention. The administration's concern about the families showed up in later years and contributed to its eventual termination. Brodey speculated that another (unnamed but not Cohen) in the administration opposed Bowen's project and pressed Cohen on this. This other was "Upset because the project was 'not looking after people'" (Brodey, personal communication, January 17, 2002).

As to Bowen's views on Scher, a letter in the archives from May 1961 gives some insight. "I was pleased to hear that Jordan Scher will be a formal discussant of my paper. He and I have a relationship that goes back to our NIMH days together. It will be good to see him again" (Bowen, 1961).

Presentation in Kansas

Bowen found a welcome respite from the tumult at NIMH when he presented a paper in Salinas, Kansas, at the joint meeting of the Kansas Association for Mental

Health, Kansas Family Life Association and the Kansas Council for Children and Youth at Kansas Wesleyan University. His presentation was aptly named "The Current Status of Man in Relation to Mental Health" (Bowen, 1954). The paper began by reviewing existing psychological understandings and then provided an original viewpoint from three areas: (1) historical perspectives that included observations of diverse forms of living things, (2) perspectives existing in the present and (3) future possibilities.

In the paper, he presented his idea that man is like lower forms of life. He introduced insights of Freud's that illuminated a new conceptual understanding of emotional illness.

> If we summarized this dimension, it might be said that it was he (Freud) who suggested that emotional disorders were not caused by physical, chemical or physiological changes in the body but instead are the result of man's relationship to man.
>
> *Bowen, 1954, 16*

No other viable theory of relationships existed and that gave a practical context to Bowen's saying that he went to NIMH to make Freudian theory more scientific. Bowen's explorations at Menninger had postulated certain replacement concepts for Freudian theory that understood the human as an evolutionary being. He said these were "to help Freudian *theory* move toward the status of an accepted science" [italics in original] (Bowen, 1988b, 351).

The paper shows insights into how Bowen was functioning. He lived in a world oriented to Freudian understanding and was trying to operate in a sphere where the human is shaped by evolution. He walked both paths. Bowen gave public notice on September 30, 1955 in the Kansas presentation that he was reviving his long-term hypothesis from the 1940s, that man was a part of all living things and therefore part of evolution.

> Beneath the social organization, man is still an animal with basic patterns shared with or evolved from lower forms. Perhaps man's need to sustain social order and to suppress any instinctual pattern that would threaten the social order would cause man to disown kinship with lower forms. Whatever the reason, an instinctual self so related to other forms of life and a social order that is much different and he has resisted serious comparison of self to lower forms.
>
> *Bowen, 1955h, 13*

Bowen posited that acceptance of such a conclusion was hindered by the potential impact on society's way of life. He established a clear "I" position in this presentation. "The presentation takes its stand squarely on the proposition that emotional problems result from man's relationships with man and that the kinds of people we are is determined largely by the character of our experiences with others" (20). If humans are under the influence of evolution, then human intellectual capacities are developing. "Man is making most rapid evolution in his power to think and as he develops this ability to think, he is making rapid strides in science, which is really the conversion of the unknown to the known" (5). He made a strong point about how changes in thinking occur. This was where Bowen himself was. His thinking was on the breakthrough idea, the family as a unit.

> The pattern of man's evolution seems to be one of a 'breakthrough'. Einstein is an example. His contribution was a new way of thinking about the world, which then permitted a rapid 'break through' on many fronts during which this way of thinking is applied to many areas. How about the world of psychological process? Primitive man conceived mental disorders as a visitation by evil spirits or that the person was bewitched. There was a period during the time of the Greeks and Romans and North African culture when mental disorder was considered sickness and treated as such. But with the advent of Christianity and the need for a more moral and ethical life, the behavior abnormality of mental disorders were considered more evil than sick and dealt with accordingly. In the 16th and 17th centuries, there began another slow turn toward seeing mental disorders as illness. Gradually … these things came to be regarded as illness and within the province of the physician.
>
> 7

Bowen introduced an evolutionary perspective on differentiation of self and foreshadowed a cross generational process that would have enormous societal impact.

> An increase in maturity is something that can be passed down through succeeding generations. It is not like physical treatment that benefits the patient and then dies in his body. If only a fraction of the small number of people in psychotherapy are able to significantly elevate their own level of functioning, then this is an overall gain to all society for future generations.
>
> *25–26*

Bowen underscored the evolutionary gains. "In this sense, emotional maturity is meant in terms of inner values and the capacity to be giving, tolerant, kind, people to whom principle is more important than social or material gain …" (25).

This presentation served as another effort for Bowen to speak meaningfully on a topic that engaged the audience to think deeply about what he was saying. It offered Bowen the opportunity for feedback without the open reactiveness present at the Clinical Center. Bowen used his experience of taking ideas outside the emotional field at NIMH as a level set for assessing his ability to present such avant-garde ideas. He engaged others to think along with him. He received very positive feedback about his presentation. In response to that letter, he noted his effort there was "to direct the ideas to a more mature level of thinking than in the usual such talk. I did that on purpose" (Bowen, 1955i). He praised the Salinas group for creating an environment that supported health, mental health and exceptionality, along with the high degree of thoughtful conversation.

Family C leaves, October

Bowen ordered a pregnancy test on the 10th for Daughter C, which came back positive on the 15th. Daughter C could no longer be in the study. Mother and daughter were told the results on the 17th. Bowen reported it to Cohen, Clinical Director, identifying the father of the pregnancy as an NIMH employee. The possibility of the family taking legal recourse against the hospital was not expected, as the family's relationship with Bowen and the project was extremely positive. Bowen reflected on the daughter's biological sensitivity to the environment. "In retrospect, we find that the pregnancy occurred when staff tension was high and the mother was staying away from the hospital" (Bowen, 1955j, 3).

Bowen's notes describe the difficulty of assessing where upset was in that household.

> This is one of those peculiar things about these mother–patient symbiotic attachments when the symptoms and the tension seem to flow back and forth from one to the other … anxiety can be seen arising in one family member and actually manifest itself in physical symptoms or emotional problems or behavior problems in the other member of the symbiosis … when mother relates … to other dependent children, the family tension can be seen popping out all up and down … through the entire family spread.
>
> *Bowen, 1955f, 1–2*

Daughter C's high sensitivity to relationship tension in the environment and her tenure in the project illustrate what later became Bowen's understanding of functional facts. A functional fact describes behavior that is an outcome of a human's functioning, not a representation of a scientific fact. He often illustrated this concept with the fact that all people dream, but the content differs from person to person. The content is not a fact, but the action of dreaming is. Bowen was thinking of systems when he commented, "I say functionally helpless because these states do not persist unless the environment plays its part in contributing to the helplessness. My research experience supports the thesis that FUNCTIONAL HELPLESSNESS CAN EXIST ONLY IN RELATION TO FUNCTIONAL OVER-HELPFULNESS" [caps in original] (Bowen, n.d.c).

At the end of October, Daughter C, accompanied by her mother, transferred back to the referring hospital (Thompson, 1955a). They initiated plans for their eventual return to the project. Mrs. C would continue seeing the social worker once her daughter transferred to the other hospital (Fisher, 1955). Daughter C's return, once the baby was born, was a possibility. Discussion about admitting the 14-year-old son also occurred (Bowen, 1955j, 6). In assuming a more motherly role in response to the needs of her daughter's new reality, Mrs. C was interpreting the situation to result from her daughter's illness (Fisher, 1955, 37). She had a surge of improved resourcefulness and a better relationship with her daughter and others this month. "Mrs. [C.] showed more progress ... than she had at any other previous time ... presented herself as less of a helpless, dependent, infantile person" (Thompson, 1955a, 40).

The involvement with this family came up in the group meeting in November where differences of opinion were expressed.

> The pregnancy was an issue in the middle of October. There was no objection about [Daughter C.] returning without the baby. On October 28, she left and a week after that [the head nurse] said the nurses were not with him [Bowen] about her returning. The only decision made without the group was about her leaving. We remember much of the wrestling with [Daughter C.] as we talk about this.
>
> *Thompson, 1955b*

Symbiosis as a relationship phenomenon, while still fitting the description in the literature, intense emotional attachment between mother and child, now, from his observations, looked more toward a biological process. Once the concept of the family unit guided the project, Bowen used more fitting terms: "emotional oneness" and "fusion." These captured, in descriptive language, the compromised function that he observed in the mother–daughter relationships. The C family did not come back to the project—another thread that an interested researcher could follow.

An intact family prepares to join the project

Bowen interviewed Mrs. D, the parent in the incoming intact family being considered for admission on October 10, 1955 (Bowen, 1955j). In choosing families for the next phase of the project, Bowen sought examples of the principle that spouses marry adequate representatives of their own parents and have an equal level of maturity. He wanted families who could live independently of their original birth families, as distinct from those who recruit a spouse for adoption into his or her family and live in proximity to the original family. A two-sibling family had advantages over larger families for observation as it had two generations where all the members are part of the interdependence (Bowen, c. 1956a). Larger families increased the number of variables. Bowen wrote of the breadth and carefulness of his thinking in selecting families compatible with the operating hypothesis for the research. "This required a detailed definition of the idea of psychological unity, the kinds of families in which this psychological unity would exist, and the specific family members who would be linked to each other in this way" (Bowen, 1959b, 3).

Mrs. D visited the ward on November 2, 1955 when the census count on the ward was ten. Of Bowen's families, one mother–daughter pair (Family B) lived in, and Daughter A came for daytime activities, her volunteer work and her appointments with Bowen. She then returned home to sleep at night. Mrs. A came for her meetings with Mrs. Basamania.

Visitors and other patients attend the group meeting

Eight others who were not the part of Bowen's core research were on the unit most of the month. This included the Genain family, along with three females under Scher's care, who were there until the 28th, when they moved to another unit. All those living on the ward attended the family-staff meetings. Often husbands, parents, children and friends of the non-research families visited and sat in on the meetings. On November 28, 1955, Scher's patients, along with the quadruplets, transferred to 2-W. Change was constant on 3-E.

There were more examples in November of nursing practices and approaches with Bowen's patients that differed from those of other doctors' patients. Families of those doctors visited patients, but not much was noted unless something unusual occurred (e.g., one patient visited with her parents in the bathroom). Nurses locked the rooms used by those patients and the quads during the days to keep them in the lounge, and the ward itself was locked. In contrast, the rooms of Bowen's families were not locked, and staff kept the unit unlocked when the other patients transferred to another unit. In the nurses' notes, other examples of contrasting practices were the hands-on care (bathing, feeding, dressing, toileting, giving directions) given to Dr. Scher's patients. On November 15 another female patient was "forcibly taken to h[ou]r with" the psychiatrist (NIMH Clinical Record, Unit Report, 1955c). Nurses forcibly removed another woman from her bed (November 15). A nurse carried one quad to the dining room (November 22). In comparison, Mrs. B had gone into town for religious services and did not show up "for her ride back from synagogue … the driver waited 10 minutes. When she called later, she was told to find her own way back" (November 12).

An open research ward

Bowen's research project was transparent, which was expressed in a willingness to receive visitors and in Bowen and staff's disposition to speak with others about the project. John Rawlings Rees, in charge of British Army psychiatry in World War II, was the first Director of the World Federation for Mental Health, London, England (Rees, John Rawlings (1890–1969), 2020) visited in November. He wrote that he found the project "very exciting" (Rees, 1955). Rees had an especial interest in the family-staff group meetings and how the background thinking developed on Bowen's project. Bowen was open to the possibility that others were doing similar things.

> If in your travels about the world you find anyone else interested in this particular kind of a family study that we are undertaking here or anyone with

a group approach similar to the one we are trying, I would like very much to know about them and to have the opportunity to exchange experiences.

Bowen, 1955c

Bowen's project attracted visitors from across the United States. Another November visitor was John M. Caldwell, Chairman, Department of Psychiatry and Neurology, Medical College of Georgia, Augusta, Georgia (Caldwell, 1955). Bowen spoke to the local community too, mentioning his project at the Washington Psychiatric and Psychoanalytic Society in November. By maintaining connections, he tracked new knowledge that could be of potential use to his theory building. And he was open to scrutiny about what he was doing.

T. P. Rees, who had visited in September, returned and sat in on a family-staff meeting on December 20. He discussed the open door policy at his facility in England (NIMH Clinical Record, Unit Report, 1955d).

Implementing the new hypothesis

To prepare for the first intact family, the research project expanded to the whole ward. Bowen wrote in the NIMH annual year-end report that his intention was to "include fathers or other close relatives" (Bowen, 1955b). He had shifted to "seeing schizophrenia as a manifestation of a distraught family that becomes focused in one individual" (3). The annual report served as an official notice of the current hypothesis. Bowen wrote that he was "redefining old concepts of schizophrenia" and alternative treatment approaches "immeasurably more effective than psychotherapy" were being used (Bowen, 1955b). And he again reiterated the family unit perspective. This updating of the change in hypothesis gave notice that those leading the Clinical Center sanctioned having entire families on the ward and the entire family in psychotherapy. Seven days later, on December 28, 1955, Family D entered the hospital (NIMH Clinical Record, Unit Report, 1955d).

Operating from a family unit orientation began when the D family arrived over two different days. The hospitalizations of the father and daughter split the family, originally from a Gulf Coast state. Mrs. D made the arrangements for transferring her husband and daughter to NIMH. A nurse from Bowen's project met the 44-year-old mother in a Midwest state to accompany her and her nearly 22-year-old daughter to the Center. They arrived on the 28th. The 52-year-old father, a transfer from another psychiatric hospital near Bethesda, joined them on the 30th. He had been treated with a course of electroshock treatment late in November for severe depression. Both daughter and father had been management problems at their previous facilities, but they each settled into the open ward without difficulty. A 14-year-old younger sister stayed at home with caretakers.

Bowen's admitting orders for this family reflected the new hypothesis.

(Mother) May take as much responsibility as she wishes regarding daughter or husband in area of privileges, meds, etc., with the privilege of asking for help in any area from the staff. Decisions about privileges (for daughter) to be made with mother and father.

NIMH Clinical Record, Unit Report, 1955d

The way the nurses recorded observations changed over time, going from individual notes on each person to one note on each family's interaction.

Mrs. D attended the remaining December unit meetings and openly spoke of the family's situation at those meetings. In early January 1956, both parents requested individual visits with the social worker. Prior to admission, the family had been told they would have individual psychotherapy. The evidence from the family's brief involvement in the group meeting showed a "higher potential for disrupting the family problem into individual problems than the mothers and patients had done. The tendency toward symbiotic involvements with the staff was much greater" (Bowen, c. 1956a). On January 9, 1956, after a series of interviews with Mrs. D, Bowen declined the request for individual therapy and elected to use a trial of family psychotherapy. By then, Basamania had held six interviews with Mrs. D and she agreed.

Bowen had observed that the other mothers, who still met individually with the social worker, attended the group meetings but did not take part. This undermined the meeting's intent. When Bowen offered the family meeting to the D's at the group meeting that day, the mother accepted it (Basamania, 1956). The reasoning behind using this treatment method was the possibility for greater progress in identifying problems coming from the openness in this meeting (Basamania, 1956, 2).

This decision came out of necessity, too. None of the therapists felt competent to see an entire family on their own. This is an interesting historical point about the move to the family group method of therapy. Trepidation prompted the therapists to use a meeting format already in place to mitigate family-staff discord and adapted it to use with the intact family. Though unorthodox, the method fit with the hypothesis. Bowen started this as a trial with the new family (Bowen, c. 1956b, 9). By June 1956, there was sufficient progress in the families to discontinue individual therapy with Families A and B, and family psychotherapy was now the only treatment method offered. The D family remained on the project until October 4, 1958. Credit goes to them for the shift to family psychotherapy.

A year of significant change

As his confidence in his reasoning that the family was a unit grew, Bowen changed the milieu and supplemented the individual treatment to better understand and observe family members. One thing that needed to be addressed immediately was the chaos caused by the anxiety transferred from family to staff and from staff to administration. One question had to be, what was this chaos obscuring?

The effort to create an integrated research design matching hypothesis and treatment began in March. Then, in his April and July presentations, Bowen endeavored to convey the magnitude of such an idea as the family as a unit. By August, he found a potential workable solution to the problem of interlocking upset between staff and family members in the new group meeting treatment. These trial-and-error approaches over the year had strengthened the idea of a family unit and the importance of fathers. In actuality, from August to December, Bowen had two treatment programs operating simultaneously. One was the effort between social worker and mother and psychiatrist with daughter, and the other was inviting family and staff to all meet as a group.

The group meeting provided the boundaries needed to contain the spread of upset between staff and family. Throughout 1955, Bowen worked toward having a better theoretical grasp of transference in a family. The new structure helped to leave it there and effectively supported the staff's contributions to a workable milieu. Bowen expected that adding fathers would strengthen the mothers' functioning and help keep sensitive issues within the family relationships.

It was in this effort that the relationship process on the unit became clear. Putting everyone together facilitated objectivity that the individual relationships did not. How anxiety transferred was observable in the group sessions. Once staff recognized it, they could address their own contributions and observing the family interaction became more accurate and predictable.

In the last quarter of the year, Bowen's planning turned to the environment that best offered opportunity for growth. Increasing the staff's ability to observe more accurately should engage them more effectively as research assistants. Minimizing ward structure should create an environment where mother or daughter found answers within self. He also wanted to end their use of hospital or ward rules as a substitute for personal decision-making.

He had carefully designed his project so that growth would occur. If there was enough strength in a family member, taking an I position should happen in a supportive and neutral environment. Taking an I position was a marker of separating. No mother and daughter twosomes had shown that strength. Bowen was stretching the research design and applying it to both staff and families in order to gather observations that better fit the new paradigm. Adding fathers and a sibling were attempts to give more strength to mothers and daughters to keep the transference within the family and to study an intact family's process. In his design, the milieu would both support and assist the innate growth in a family member, leading them to action for greater independence.

> It is believed that maximum freedom from external control will provide a clearer picture of the true character of the basic relationship and more opportunity for activation of the evolutionary growth of the arrested symbiosis.
>
> Ward treatment structure: A treatment environment designed to give as much freedom as possible in the intra-unit relationships and as much psychological support as possible to provide the maximum in expression of forces

within the unit, a maximum opportunity to break out of the fixedness of the symbiosis toward maturation.

No author, c. 1955

Stepping into the unknown

The year was a steep learning curve for Bowen and staff. The last months showed both progress in and obstacles to Bowen's project. Beginning in August, pressure came from the administration to produce some evidence of the work being accomplished. Whether or not intentional, the presentation in August at Chestnut Lodge was timed with the publication of a major paper on milieu by other researchers. Showcasing one of their own projects meant the Clinical Center could measure up to the field's trendsetters. It is ironic that Bowen's project was viewed as chaotic and disordered, yet it could represent the best the Center offered.

Indicators of larger processes within the institution could also affect Bowen's research efforts. An example is seen in Bowen's note of December 21, 1955, the date of his annual report.

> As the staff prepared for the project's continuance with admissions scheduled for the first father, mother, impaired offspring family at the end of the month, … the Executive Committee suspended further admissions to the project, appointed [a] committee, and permitted the [D] family to be admitted provided their admission was understood to be temporary and with the understanding that they might be asked to leave.
>
> *Bowen, c. 1956c*

The family arrived a week after Bowen received this memo. What caused this action by the Executive Committee is unknown but Bowen saw it as a potential obstacle.

> I personally am beginning to get a definite although still very vague feeling that schizophrenia may not be completely reversible in a government setting and that the highest goal that might be possible in this setting is a remission or amelioration goal.
>
> *Bowen, c. 1956c*

In a world operating in psychoanalytic theory, Bowen kept a presence there while having an orientation to the world of the family as a natural system. Was the Executive Committee sensing his new orientation to the family as a natural system and having doubts about what he was doing? Many questions are left unanswered.

Chapter 15 looks at the Genain family, of interest to Bowen's theory building but not part of Bowen's formal research. Observations of this family ran parallel to his research in his first year at NIMH. But there is direct evidence in the NIMH records that observations of the sisters' functioning contributed to the differentiation of self-scale. What a year 1955 was.

References*

*Unless otherwise specified, the works of Murray Bowen included in this chapter are from the Murray Bowen Papers. 1951–2004. Located in: Modern Manuscripts Collection, History of Medicine Division, National Library of Medicine, Bethesda, MD. The Accession, Box and Folder information differs for each one and that is included here.

Basamania, B. (1956). [Summary of casework contacts with (Mrs. D name redacted) from admission on 12/28/55 through 1/9/56]. L. Murray Bowen papers, National Library of Medicine, History of Medicine Division. (Acc. 2006-003, Box 3, Folder Clinical record copies), Bethesda, MD.

Bowen, M. (n.d.a). [Letter]. (2005-055, Box 3, Folder M. Bowen's working papers for an insurance claim (name redacted)).

Bowen, M. (n.d.b). [Draft]. (Acc. 2007-073, Box 3, Working Papers from NIMH project re: Schizophrenia).

Bowen, M. (n.d.c). [Letter to editor]. (2007-012, Box 3, Folder Letter to editor).

Bowen, M. (1954). [The current status of man in relation to mental health]. (Acc. 2006-003, Box 5, Folder Mtg. Notes October 1954).

Bowen, M. (1955a). [NIMH Bulletin board notice of a pending meeting]. (2006-003, Box 7 Folder Misc. papers).

Bowen, M. (1955b). [Annual report]. (Acc. 2006-003, Box 4, Folder Analysis of NIH program activities description sheet).

Bowen, M. (1955c). [Letter to John Rawlings Rees on December 29, 1955]. (Acc. 2007-073, Box 3, Folder Clinical notes on patient families in NIMH project efforts).

Bowen, M. (1955d). [Clinical presentation to clinical investigations conference, (A). family]. (Acc. 2006-003, Box 4, Folder Clinical staff presentations August).

Bowen, M. (1955e). [Draft of letter to Dr. Robert Heath]. (Acc. 2007-073 Box 3, Folder Former patient family [name redacted]).

Bowen, M. (1955f). [Memo to Dr. Robert Cohen, Report of the clinical course of Miss (C), October 27]. (Acc. 2006-003, Box 3, Folder name redacted, Daughter C).

Bowen, M. (1955g). [Letter to a colleague]. (Acc. 2006-003, Box 6, NIMH correspondence H through P).

Bowen, M. (1955h). [The current status of man in relation to mental health]. (Acc. 2006-003 Box 5, Mtg. Notes October 1954).

Bowen, M. (1955i). [Letter to director of the Kansas State Board of Health for Mental Hygiene, October 13, 1955]. (Acc. 2006-003, Box 6, Folder Meeting attendance).

Bowen, M. (1955j). [Appointment calendar]. Copy in possession of the Bowen family, Williamsburg, VA.

Bowen, M. (c. 1956a). [Draft. A psychological concept of schizophrenia]. (Acc. 2007-073, Box 3, Folder Working papers/schizophrenia: A family concept of schizophrenia).

Bowen, M. (c. 1956b). [Draft. Psychotherapeutic treatment of the family as a unit]. (Acc. 2006-003, Box 2, Folder drafts).

Bowen, M. (c. 1956c). [Issue of executive committee suspending admissions on Dec. 21, 1955]. (Acc. 2007-073, Box 3, Folder Clinical notes on patient families in NIMH project).

Bowen, M. (1956a). [Admission note on the (D) family]. (Acc. 2006-003, Box 3, Folder Clin. record copies).

Bowen, M. (1956b). [Letter to Leslie Osborn, July 27, 1956]. (Acc. 2006-003, Box 7, Folder Osborn, Leslie A., Wisc. Psychiatric Inst.).

Bowen, M. (1956c). [Clinical record—Doctor's progress notes, May 7, 1956]. (Acc. 2006-003, Box 3, (Son C. [name redacted])).

Bowen, M. (c. 1957). [A psychological concept of schizophrenia]. (Acc. 2007-073, Box 3, Folder Working papers/schizophrenia: A family concept of schizophrenia).

Bowen, M. (1958). [Psychotherapeutic treatment of the family as a unit]. (Acc. 2006-003, Box 2, Folder drafts).

Bowen, M. (1959a). [Letter to Rudolph Ekstein on October 12, 1959]. (Acc. 2004-013, Box 4, Folder PSA Institute).

Bowen, M. (1959b). [Draft. Family psychotherapy]. (Acc. 2006-003, Box 2 Folder Rough drafts—Ortho).

Bowen, M. (c. 1960). [Draft. Family study clinical research project]. (Acc. 2007-073, Box 3, Folder Family study clinical research project).

Bowen, M. (1961). [Letter on May 3, 1961 to Medical Director, Forest Psychiatric Hospital]. (Acc. 2004-043, Box 5, Folder Dr. Bowen's paper).

Bowen, M. (c. 1964). [Intrafamily dynamics in emotional illness]. (Acc. 2004-013, Box 2, Folder Ggtwn conf. Jan 1964).

Bowen, M. (1975). Family therapy after twenty years. Chapter 14 in *Family therapy in clinical practice*. New York, NY: Jason Aronson, 1978.

Bowen, M. (1985). [Letter, January 22, 1985 to Mary Bourne]. (Acc. 2003-044, Box 6, folder Minneapolis, Minn. – July 28–29, 1986).

Bowen, M. (1986). [Letter to Bob Dysinger, October 15, 1986]. (2007-012, Box 2, Partial drafts).

Bowen, M. (1988a). [Attachment to letter to a student in the post-graduate program at Georgetown family center, October 8, 1988]. (Acc. 2004-043, Box 5, Folder Letters from Dr. Bowen).

Bowen, M. (1988b). Chapter "Epilogue, An odyssey toward science". In *Family evaluation: An approach based on Bowen theory*. New York, NY: W. W. Norton & Company.

Bowen, M., Dysinger, R., Brodey, W., & Basamania, B. (1957). [Study and treatment of five hospitalized family groups each with a psychotic member]. (Acc. 2006-003, Box 6, Folder American Orthopsychiatric Ass'n].

Caldwell, J. M. (1955). [Letter to Dr. Lucius M. Bowen, November 21, 1955). (Acc. 2006-003, Box 6, Folder NIMH correspondence A through G).

Fisher, T. (1955). [Summary of case work help to Mrs. (C) during the month of October 1955]. L. Murray Bowen papers, National Library of Medicine, History of Medicine Division. (Acc. 2006-003, Box 3, Folder Clin. record copies), Bethesda, MD.

Gilliland, E. G. (1957). In memoriam… A tribute to Louis M. Cholden, M.D. Mountain Scholar. Retrieved November 27, 2020, from: https://mountainscholar.org/bitstream/handle.10217/184925/MMTA01201_06.pdf.

Jones, M., & Rapoport, D. (1955). Administrative and social psychiatry. *The Lancet, 266*(6886), 386–388. https://doi.org/10.1016/S0140-6736(55)92368-0.

NIMH Clinical Record, Unit Report. (1955a). [NIMH Clinical Record, August 1955]. L. Murray Bowen papers, National Library of Medicine, History of Medicine Division. (Acc. 2006-003, Box. 1, August 1955), Bethesda, MD.

NIMH Clinical Record, Unit Report. (1955b). [NIMH Clinical Record, September 1955]. L. Murray Bowen papers, National Library of Medicine, History of Medicine Division. (Acc. 2006-003, Box 1, Folder September 1955), Bethesda, MD.

NIMH Clinical Record, Unit Report. (1955c). [NIMH Clinical Record, November 1955].
L. Murray Bowen papers, National Library of Medicine, History of Medicine Division.
(Acc. 2006-003, Box 1, Folder November 1955), Bethesda, MD.

NIMH Clinical Record, Unit Report. (1955d). [NIMH Clinical Record, December 1955].
L. Murray Bowen papers, National Library of Medicine, History of Medicine Division.
(Acc. 2006-003, Box. 1, Folder December 1955), Bethesda, MD.

No author. (c. 1955). [Draft of annual report]. (Acc. 2007-073, Box 6, Folder Clinical Notes
on Patient Families in NIMH Project).

Rees, J. R. (1955). [Letter to Dr. Bowen on December 31, 1955]. L. Murray Bowen papers,
National Library of Medicine, History of Medicine Division. (Acc. 2007-12, Box 4,
Folder Misc.), Bethesda, MD.

Rees, John Rawlings (1890–1969). (2020). Retrieved January 23, 2021, from encyclopedia.
com: www.encyclopedia.com/psychology/dictionaries-thesauruses-pictures-and-press-
releases/rees-john-rawlings-1890-1969.

Referring physician (Name redacted). (1955). [Letter to Dr. Bowen on August 30, 1955].
L. Murray Bowen papers, National Library of Medicine, History of Medicine Division.
(Acc. 2006-003, Box 3, Folder Name redacted correspondence), Bethesda, MD.

Thompson, C. (1955a). [L. Murray Bowen papers, National Library of Medicine, History of
Medicine Division]. (Acc. 2006-003, Box 4, Folder Clare's mtg. notes & research charts),
Bethesda, MD.

Thompson, C. (1955b). [Chart October 21, 1955 to October 28, 1955]. L. Murray Bowen
papers, National Library of Medicine, History of Medicine Division. (Acc. 2006-003, Box
4, Clare's mtg. notes & research charts), Bethesda, MD.

15

AN IMPORTANT NON-RESEARCH FAMILY

The Genain family

The Genains were well known in psychiatric circles for having four monozygotic schizophrenic daughters. Their name, Genain, a pseudonym meaning "dire birth" (Hahn, 2019), was given to them by David Rosenthal, the psychologist genetics researcher and Chair of the Research Committee, who studied them at the National Institute of Mental Health (NIMH). There was significant excitement to have the quadruplets at NIMH; the administration expected substantial exploration possibilities along with the reputation that would accompany them. Over their time at NIMH, five laboratories collaborated on studies of these young women to understand the biological/genetic factors of the origin and development of schizophrenia (Farreras, Hannaway, & Harden, 2004).

My primary reason for discussing the Genains is the opportunity they provided Bowen to observe an intact family at the time he was forming his understanding of a family as a single entity. What was observable with Mr. and Mrs. Genain was pertinent to how Bowen's ideas were forming throughout 1955. The quad sisters and their parents, hospitalized on Bowen's ward at least nine times that year, when beds were unavailable elsewhere in the hospital, added a different dimension to Bowen's family studies and within the institution. The families Bowen admitted to his research project had one mother and one child. Now, there were four sisters, age 24, with both parents living with them. This gave Bowen a unique opportunity to interact with and observe the sisters with their parents on the ward with his research families. The notes that Bowen recorded about this family give insight into his approaches to the data that would contribute to the development of his concepts in the 1960s.

The Genains also give insight into the conflation with staff at NIMH as to who is seen to be responsible for this family. Bowen and Mrs. Basamania, who at

DOI: 10.4324/9781003027287-16

that time was not working on Bowen's project, were both involved in the family's transfer to NIMH (Robert Cohen Interview with Catherine Rakow, 2002), though placement on the unit where Cholden and Bowen had their projects was an administrative decision. While the family was not one of Bowen's research families, others at NIMH understood Bowen to be the one in charge of the Genain family. Most likely, the administration's vagueness as to who held therapeutic responsibility for this family contributed to confusion in the staff and among other families on the ward. The same nurses who worked on his study also served the Genain family's health care needs. The families on the ward all mingled in the common areas. Friendships formed between the daughters of the families on the ward. And the conflation of responsibilities created tension between Bowen and the administration.

Bowen observed the same patterns in the Genain family interactions that he had seen in the Menninger research families. There was a lack of definition between the parents. The father operated as an extension of the mother, not as a person in his own right. Mother's dominance in the husband–wife interdependent relationship only gave the appearance that she was a separate self. The parents' fusion was clearly a part of the family problem, as much as it was in the Dutch family.

A research committee forms to define a path forward with the Genains

It was seven months after their first placement on the unit that an effort began to develop a research direction with the sisters. Bowen was on the committee of researchers that met weekly beginning in August to discuss this family. He was one of four therapists working with the sisters individually since the family's arrival. Rosenthal chaired the meetings, and Lewis Hill was the consultant to this group. Bowen was the clinician for the sister labeled "N"ora. Rosenthal, in his 1963 book, labeled the sisters Nora, Iris, Myra, Hester so their initials were NIMH (Rosenthal, 1963).

In November, Bowen transferred sister N to another psychiatrist. It is unknown whether this transfer was his idea or that of NIMH.

At the initial meeting of the Research Committee, Hill described the approach to the family since January as matching the approach on Bowen's project. He said it was a

> neutral position to permit the conflicts to arise between parents and the patient and for once the patient doesn't arbitrarily [feel] compelled by hospital rules to do what the parents think … to put the parents on the spot for a change.
>
> *NIMH Clinical Record, 1955*

Bowen was aware that others assumed he was "held responsible for these people" and asked Jenkins, Clinical Administrator and committee member, to take this family "over immediately" (NIMH Clinical Record, 1955). It was Bowen's opinion that the amorphous situation with this family had not been challenged "because it gets the parents off of everybody else's back so who's going to complain?" (Bowen, 1955a). He also suggested that a gap in knowledge had resulted in the institution's acting as Mrs. Genain's servant (Bowen, 1955b).

No theory for understanding or treating a family group then existed to guide the Committee. A climate of tension had developed around defining operating principles for the study of the Genain family, and the difference between operating with and without a theoretical approach is never clearer than in the transcripts of the committee's meetings. Discussing a research approach for the four sisters, Bowen suggested first defining an attitude toward the patients on which a platform could be built to work with the patients and family (Bowen, 1955a). His attitude was that the sisters' symptom of schizophrenia was an overworked strength, a concept tracing back to his Menninger work, and he described an attitude of interest, openness and listening for maturity as fundamental to relating to the patients and the family.

Bowen used these sessions to put his ideas out there for others to discuss. The records of these meetings show that he openly spoke about these ideas to his peers. He offered his operating principle to involve the patient in decision-making: "Here's something else … I'm against with schizophrenics, making any kind of a move without their participation in it" (Bowen, 1955b). Bowen suggested establishing each sister and clinician as autonomous entities and noted the expected benefits of this attitude for the sisters and the administration. Another operating principle he suggested was to relate to the "adult" in a sister and to recognize the "infantile" side but not to relate to it.

The research meetings allowed Bowen not only to speak about his own approach but also to interact with his peers without using Freudian thinking or resorting to the construct of id-ego-superego. Bowen explained that the simplest way to understand human behavior was to pay attention to what people actually do and say; using id, ego and superego to interpret human behavior was not required.

Bowen described the sisters as one with the mother. Their oneness generated instability between them and inhibited individuation. Losing definition between people was part of the family dilemma. He said that he told his patient/sister Nora it may be possible to emerge from the family stuck-togetherness in his sessions with her.

Bowen spoke with the other researchers about an individual's capacity to separate feeling and reasoning. Being able to separate them was the important point. Hill asked Bowen if he assumed that "Somewhere is the need … to have a good relationship with the good mother that is strong enough that you can rely on it" (Bowen, 1955c, 1). Bowen replied, "Right, then on that I'll bank the whole thing …. It lives or dies on that." The "good mother" assists the offspring in maturation; even if that is not innate, a mother can learn how to do it. It is the functioning position, how it is carried out, and the stability of the father and mother's relationship that makes the difference. Keeping the separation process within the family's boundaries builds in durability. And leaving the process between family members put Bowen in the position of consultant to the individual efforts of sister pseudo named Nora. He told the committee that this was his attempt to keep any transference at a neurotic level.

By the third meeting, the topic turned to supervision and to desirable characteristics in a supervisor. Hill stressed the importance of knowing self in the relationship and how that could produce responses unique to the therapist and patient. Bowen described wanting openness in a relationship with a supervisor.

> What happens with the usual supervisor, is it gets to be a kind of a business of him trying to sell a way of doing things, and either you are buying it or not. What I have sought in supervision … is somebody to point out … what I am doing in my way … you've got to have … somebody else help you see what you are doing
>
> *Bowen, 1955d*

Subsequently, Bowen showed the committee a diagram he had created to show his understanding of the functioning among the sisters and their mother (see Figure 15.1). In Figure 15.1, the names used, for confidentiality purposes, are those Rosenthal assigned in his 1963 book to identify the sisters (Rosenthal, 1963).

Bowen explained that he was looking at the sisters' relationships with each other and to the mother in terms of their functioning. These notes reveal Bowen's inner thoughts on functioning and differentiation. It would appear that one sister was dominant within a set. Bowen attributed the dominance in the pairs to the mother's greater recognition of one daughter, not to some inner strength of the daughter. Within a dominant/adaptive pair, the dominance is an artifact of recognition by a third, important person to the twosome (Bowen, 1955b). He described the quadruplets' functioning as two sets of two. Within and between each set was a shifting balance in functioning. When the balance was seriously out of kilter, as shown by the second scale on the diagram, then the shifts were only in relation to the mother.

Understanding dominance as a relationship process is interesting in the broader context of the Clinical Center. Records show that the administration's anxiety about Bowen's project was rising at the time the model of the family group meeting was forming (August 1955). Contributing factors to the administration's apprehension were the increasing responsibility given to the families in Bowen's project and the perceived equalizing of everyone on the ward. Bowen described the shifting alliance among the hospital, the family and the clinician to make his case. For example, in response to its increasing anxiety, the hospital administration was acting similarly to the families in Bowen's project by keeping the quadruplets in the dark about administrative decisions impinging on them. From the vantage point of some 70 years later, these records bring another insight. Bowen's project and his presentations outside of NIMH demonstrate that he went beyond conventional thinking. While it would have been difficult at the time to interpret the

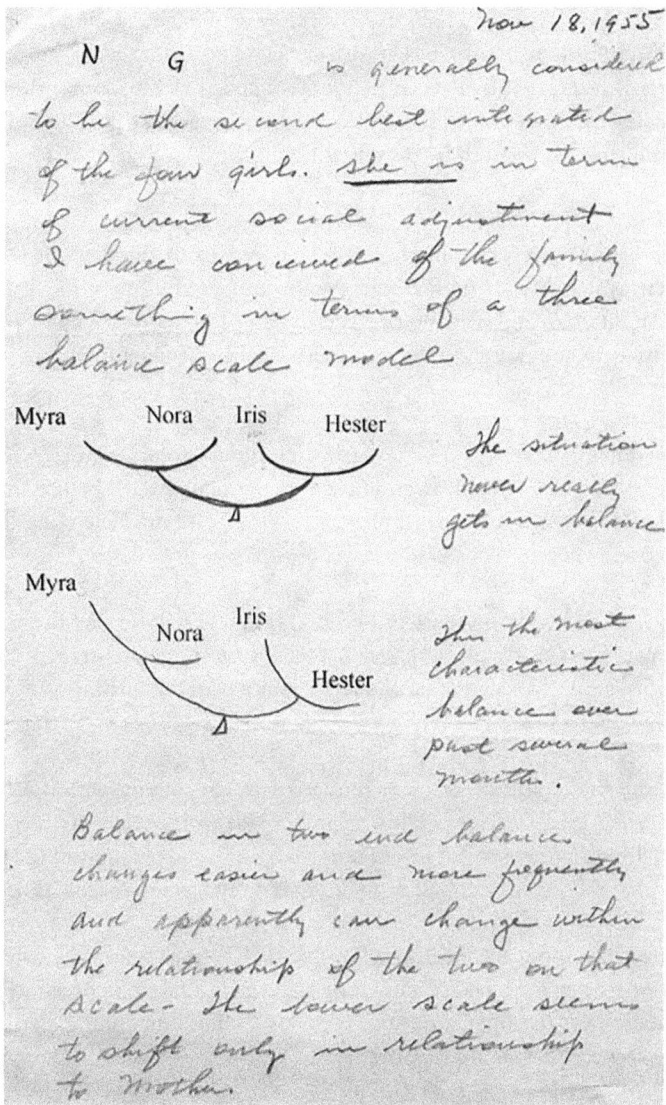

FIGURE 15.1 Bowen's handwritten notes with two diagrams, 1955

Source: Reprinted courtesy of the National Library of Medicine, History of Medicine Division, Bethesda, MD. Bowen, M. (1955c). [Drawing]. (Acc. 2006-003, Box 3, Folder [Name redacted]).

administration's anxiety as pushback on that point, it seems reasonable now to suggest it.

These brainstorming meetings for the Quad project reveal varied viewpoints. An attendee responded to Bowen's thinking: "Now from a research angle, one could easily set this up as a pair in which the supervisor would see the symbiotic relationship, the sort of scale relationship that Murray has talked about many times between those two" (Bowen, 1955d). Bowen responded, "I tried to work out some kind of a thing which would put the mother as the fulcrum of the overall scale ..." (Bowen, 1955d). Hill gave his opinion by paraphrasing the work of psychiatrist Harry Stack Sullivan.

> There is no society of four, there can be two persons, there can be three, that's as far as it goes as a unit. Because with four persons, they inevitably break down into two and two and one and three or something of that sort ... it is a pretty firm conviction of his, you don't see four people functioning as a unit.
>
> *Hill, 1955*

As discussed in Chapter XIV, a new ward administrator took over the 3-E ward on September 8. Two days later, Bowen noted that the value placed by the administration on the project with the quadruplets was to the point of challenging the family project's continuation.

> I was asked to put my project in mothballs .... I believe that the issue came up over the Quads and I believe it belongs more to [the] tremendous value we place on the Quads, that they have value over other things.
>
> *Bowen, 1955e*

This was the first evidence of overt tension with the administration. While they recognized that long-term research on families was worthwhile, they would put a hold on Bowen's family project in order to study this unique family. It did not seem to matter that the institution had not yet figured out a research direction with the siblings.

The meetings about what to do with the quads foreshadowed the eventual termination of Bowen's project, precipitating the administration's choice to focus on schizophrenia over the opportunity to focus on family functioning. Perhaps some future researcher will take up that history.

Extensive history on this family

While writing "Treatment of Family Groups with a Schizophrenic Member" for the March 1957 American Orthopsychiatric Association (AOA) annual meeting, Bowen interviewed the Genain family and created diagrams of the family's history.[1] His explicit purpose for doing these diagrams is unknown, but the annual meeting may well have provided the motivation, as it included a section on family research. Bowen dated his exploration of the extended family idea to 1956–1957 (Bowen, 1963) and his use of "family psychotherapy" (possibly the first use of that phrase) to this meeting (Bowen, 1976, 351).

Bowen's diagrams of the Genain family (meant to help explain the concepts that he was trying to get across) were of the nuclear family, the paternal extended family and the maternal extended family—four generations. A handwritten history accompanied the drawings. He was using the diagram and history to provide data for his questions. What is occurring when one person does better from another? How is a family member important to a marriage? How are tensions in one family member acted on in another family member? What is the togetherness balance in the family? What are the functioning differences among siblings? And what patterns repeated from extended family to nuclear families? Bowen used diagramming as a search for evidence of varying levels of attachment in one generation between offspring and mother; of increased attachment and dependency between parent and child down specific branches; of repeating patterns in particular branches; and of cutoff from family in a particular branch.

It is time to return to Bowen's odyssey. In Chapter 16, I review the question at the heart of this book: Where did the theory originate?

Note

1 This paper is Chapter 1 in Bowen's book, *Family Therapy in Clinical Practice*, Jason Aronson: New York, NY, 1978. The diagrams are in the archives but not in the book.

References*

*Unless otherwise specified, the works of Murray Bowen included in this chapter are from the Murray Bowen Papers. 1951–2004. Located in: Modern Manuscripts Collection, History of Medicine Division, National Library of Medicine, Bethesda, MD. The Accession, Box and Folder information differs for each one and that is included here.

Bowen, M. (1955a). [Transcript. Initial meeting of therapists for the four (Genain) girls with Dr. Lewis B. Hill, consultant on August 3, 1955]. (Acc. 2006-003, Box 3, Folder Name redacted quadruplets therapists meeting).

Bowen, M. (1955b). [Second meeting of the (quadruplet) therapists with Dr. Hill, consultant on August 10, 1955]. (2006-003, Box 3, Folder [quadruplets] therapists meeting).

Bowen, M. (1955c). [Drawing]. (Acc. 2006-003, Box 3, Folder [Name redacted]).

Bowen, M. (1955d). [Transcript. Third meeting of the quad therapists with Dr. Hill, consultant on August 17, 1955]. (2006-003, Box 3, [quadruplets] Therapists Meeting).

Bowen, M. (1955e). [Transcript. Seventh meeting of the (name redacted) therapists with Dr. Hill, consultant on September 14, 1955]. (2006-003, Box 3, Folder [Name redacted] therapists meeting).

Bowen, M. (1963). [Letter on July 17, 1963]. (Acc. 2011-015, Box 4, Folder Patient inquiries).

Bowen, M. (1976). Theory in the practice of psychotherapy. Chapter 16 in *Family therapy in clinical practice*, New York, NY: Jason Aronson, 1978.

Farreras, I., Hannaway, C., & Harden, V. (Eds). (2004). *Mind, brain, body, and behavior: The foundations of neuroscience and behavioral research at the National Institutes of Health.* Amsterdam: IOS Press.

Hahn, P. (2019, October 17). Hereditary madness? The Genain Sisters' tragic story. *Mad in America*. Retrieved November 20, 2020, from www.madinamerica.com/2019/10/genain-sisters-hereditary-madness.

Hill, L. (1955). [Transcript. "Third Meeting of the quad therapists with Dr. Hill, consultant", August 17, 1955]. (Acc. 2006-003, Box 3, Folder [name redacted] therapists' meetings).

NIMH Clinical Record (1955). [Transcript. "Initial meeting of therapists for the four [Genain] girls with Dr. Lewis B. Hill, consultant]. (Acc. 2006-003, Box 3, Folder [Name redacted] therapists' meetings).

Robert Cohen Interview with Catherine Rakow. (2002, June 12). *The Murray Bowen archives project*. Retrieved September 2017, from http://murraybowenarchives.org/oral-history-11-Cohen.html.

Rosenthal, D. (Ed.) (1963). *The Genain quadruplets: A case study and theoretical analysis of heredity and environment in schizophrenia*. New York: Basic Books.

16

ORIGIN OF BOWEN'S THEORY

Bowen's cross generational study at National Institute of Mental Health (NIMH) is well known today among the students of the theory. In psychiatry, his study is recognized for its contribution to family systems psychotherapy.

Murray Bowen was a medical doctor, psychiatrist and researcher who devoted his professional life searching for and planning a theory of human functioning consistent with the disciplines of contemporary science, biology in particular. He used a practice of intensive literature review and qualitative, observational research to study individuals in their family contexts. Based on those observations, he then envisaged a set of commonly seen dynamics. Those dynamics or concepts formed the basis of his theory of the family as a naturally occurring, regenerating system.

The definition of a natural system fits with a definition of family—regularly interacting or interdependent elements that form a unified whole. Each element functions independently and in congruence with the other elements, which have or tend to an equilibrium under the influence of related forces (gravity, magnetism and individuality/togetherness). It is observable in nature. It forms in response to other elements in nature. Bowen's theory, which came to be known as family systems theory, takes these elements and formalizes them into a unified whole. Today, many mental health practitioners, educators, organizational consultants and leaders in communities of religious faith use Bowen's family systems theory as an assist in addressing troublesome entanglements or advancing individual growth.

This book looked at the research process Bowen used to develop his theory, particularly during the first ten seminal years of his professional life as a psychiatrist. His approach offers us a model for continued research and theorizing in this large and important area of interest: human life. Bowen's model is of value not only for professionals but also for individuals seeking to understand their own lives and more intentionally direct their efforts to fulfill their aspirational values and

DOI: 10.4324/9781003027287-17

goals—for their own advantage and for the benefit of their families, organizations, communities and generations to come.

When people speak of theories, they are often referring to formal constructs in the different fields of science. But creating a "theory" about how the world of the human works is something every human does from birth with greater and lesser degrees of complexity. We each attempt, through trial and error, to organize the data we receive through our senses. Organizing and patterning the details of life involves interaction and dialogue with others of our kind. Making sense can be a collaborative process that helps us refine our theories and thus more accurately guides our contact with the world. We live with an urgency to make sense of the reality of our world. We need ideas and concepts for life that will help us adjust to that reality. And we can consider those ideas and concepts as theories—collections of facts relevant to each other.

Science is a process of continual change, revolutionarily so at times, as physicist, historian and philosopher of science, Kuhn (1970) has observed. Like Kuhn, I believe it is a process that must remain open to revision lest we fall into the problematic position of dogmatizing or clinging to our former, no-longer-valid conclusions about life, especially human life. I believe Bowen's work and personal history, where he continually pursued and revised his best grasp of reality, exemplified that openness and that determination.

In this sense, developing a theory is not a step-by-step process. Rather, it is a search: putting clues together; seeing if they hold up; giving up familiar ideas; searching some more and finding a new clue; putting one idea with another; all with a willingness to learn, learn and learn, and then trying to talk and write about it. In Bowen's archive of research materials, we can see that he lived this steady effort to grapple with the often complex and confounding progression of gathering and making sense of his research data.

Murray Bowen decided to scientifically study humans and developed a theory of the human that is more profound and comprehensive than many of the more "scholarly" theories prevalent during his early years of training, and even more than those in common use throughout his professional life. Because of his determined effort over the course of at least 45 years (from 1945 until his death in 1990), he left not only the theory behind but also an archive of his records that gives us an unparalleled view into his method of researching and theory construction. In the details of Bowen's particular history, we may derive a set of principles that describes the more universal manner of human theorizing, testing and revising. I believe we have much to learn from his example.

As I tried to show in this book, Bowen's story exemplifies the actual practice by which theories are made and remade, in effect offering us a case illustration of how science itself progresses, how wisdom is generated and how it moves to future generations. In fact, my process of creating this book—the many years of archival research, reflection and distillation—has been my own "odyssey," to borrow the descriptor Bowen used about his own research efforts over those many years.

This book has looked at ten years of Bowen's journey that began with an immersion in learning and practicing Freudian theory. His path then led to research at both Menninger and NIMH, culminating in a revolutionary realization about families. There was enough substance in his work to form an original theory. From 1949 through 1955, it was a course of "spurts and stops" (Bowen, 1986). Following admission of the first intact research family to his project, Bowen "KNEW that somehow, somewhere, in all of this would be a new and different theory of human development ... involved in evolution and natural systems but THAT THINKING INFLUENCED EVERY IDEA AND CONCEPT THAT FOLLOWED" [caps in original] (Bowen, c. 1986). This quote and his questioning of observations, searching and seeking of confirmatory facts hint at the uncertainty deep within Bowen in the years before he admitted the first intact family to his project. It is easier today to infer the confidence in his choices than it was for Bowen to do so at the moment.

The question posed at the start of this book was where Bowen's theory originated. I have attempted to document the conclusion that it originated during Bowen's years at the Menninger Foundation. The discovery at NIMH of family as a unit was a catalyst, directing the approach to hospitalizing intact families for authentication. So the answer is the theory originated at Menninger's and became cohesive at NIMH.

In 2021, I found a confirmation of that in Bowen's writings. At the 1987 symposium, he said,

> Now the theory was well developed in Kansas, down to specific details, every little detail except for one. Then I got to NIH .... And suddenly the years in Kansas made it possible for me to see things in families that nobody else in family saw .... I conceptualized family as an evolutionary form of life.
>
> *Bowen, 1987a, 18*

That quote works for my purposes. But it does not capture the processes in Bowen's mind in those early months at NIMH. Bowen did not have a specific research design when he arrived at NIMH. When he planned the design, "The treatment program had been organized with the implicit assumption that the symbiotic union involved chiefly two people" (Bowen, 1957). In the months of creating his project design, he also saw the Dutch family clinically as out-patients. The meanings of the observations he recorded are obvious now in retrospect. The interdependence in a family went beyond two people. As I thought about this, it seemed he had layers of awareness. One addressed the short-term aim of getting a project up and running; another layer understood what he observed in light of a long-term goal of defining the facts of human functioning. When his research at NIMH repeated and confirmed the findings he had seen in the out-patient family, a merging of those layers of knowledge occurred and he had his Aha! moment. The family is a unit.

When the Clinical Center hired Bowen, it was seeking practicing clinicians with a solid background in applying Freudian theory innovatively. He was, by then, well on his way to an original theory in that moment of his life and studies. Bowen had positioned himself on the periphery of psychoanalysis; he defined it "just outside the conceptual boundaries of medicine" (Bowen, 1995, 22). It is helpful to look at Bowen's own written thoughts about trying to reach toward science and extend Freudian theory during his NIMH project. Bowen wrote this around 1956 regarding the NIMH hypothesis; the paper, discovered among the NIMH materials, was published in 1995.

> Psychoanalysis is seen as … occupying a bridge position between the medical and non-medical positions. Freud … recognized the conceptual disharmony between physical and psychic functioning … chose to bridge the gap by the use of a medical conceptual model to bring the new order of events within the boundary of medicine. The union has never been a smooth one. This project has its basic orientation in that part of psychoanalysis which is just outside the conceptual boundaries of medicine … it is proposed to 'reach toward' the medical and biological disciplines in an effort to make conceptual contact … by conceptualizing certain biological events from a psychological viewpoint. The term 'biological' is used in a sense distinctly different from physiological … it refers to life processes such as birth, growth, reproduction and death, which are common to all protoplasmic life. The psychological thinking attempts to view man as a biological and instinctual being. Theoretical formulations of father as a social, ethical, moral and intellectual being are excluded if they are in conflict with the biological viewpoint.
>
> *Bowen, 1995, 21–22*

Saying he wanted to explore a theory to replace Freudian theory at the outset would have likely ruled out a candidacy at NIMH. Timing was not yet right to take that step: "Social innovation is often a discovery, even to those involved" (Westley, Zimmerman, & Patton, 2006, 225).

It is notable that while Bowen was confident he was on to an alternative understanding, he did not put the idea that the family was a unit into words until he had evidence in multiple family pairs on his ward. He was a cautious researcher. The *Time* magazine ad, published writings of others using the phrase "family as a unit," and at this point late 1954, early 1955, Bowen had the idea but not the data that would place the family within evolution.

Fundamental change

Theories focused on the individual rather than on the family as a system are still the operative theories of understanding living things today. Bowen's work has contributed to a future theory, that of systems of living things.

The human, as a scientific being, awaits documentation by others. Bowen reflected on his effort. "Since the late 1940s I had a background hunch that it [evolution] somehow relates emotional illness to that part of man he shares with the lower forms of life, but I could find no way to implement this idea" (Bowen, 1976, 394). It was in the mid-1950s that he dedicated his life to that purpose.

As this book has demonstrated, it is possible to follow Bowen's progression as he gathered data about human behavior. A vital change began with Bowen's literature search during his residency. That practice continued throughout his life. He did a broad and thorough analysis of studies across disciplines. He found differing perspectives on the use of facts in animate and inanimate theories. This search led him to his own viewpoint of the essential assumptions required for understanding human behavior. Evolution, with its base in a systems understanding, met his criteria for an essential assumption.

The theory he formed at Menninger was accurate in its application there. Applying it with actual family members led to recognizing the missing link in his theory, the family as a unit, and announcing it in 1955. Now his hunch of the human as similar to other living things was a testable hypothesis, researchable using fundamentals of living systems.

Bowen had the lofty goal of understanding human behavior, with an explanation grounded in facts coherent with nature, at least as far back as 1946; until he concluded the family was an analyzable unit based in evolutionary biology, his orientation remained a psychological one.

The retrospective account presented in this book may give an impression of an orderliness that is misleading. It is important to remember when Bowen wrote that his intention in his research at Menninger was to advance Freudian theory, there was no other accepted theory. To state otherwise in moving to NIMH in 1954, lacking solid data, would have been irresponsible.

Thirty years later, he recalled he had enough facts to form his new theory at Menninger, before he came to NIMH. He stated, "The more scientific theory was defined in detail in Kansas" (1988). By the 1980s, being well established as an original thinker and a pioneer in studying the family, he could speak readily about pacing from one idea to another. That made his theory building progressive and orderly. But questioning, then replacing what one has learned while still operating in a world that thinks with the old knowledge, surely had its challenging, disturbing moments.

Bowen left Menninger for NIMH with an awareness of a two-person process. Schizophrenia developed within a parent–child relationship dependent on certain relationship conditions in that relationship (the Menninger regression experimentation showed this). The efforts at Menninger had shown that the actual mother's presence would allow the child a learning opportunity. A parent's presence made a difference in treatment outcome. At NIMH, it was a pure new search for facts using extreme symptoms, namely schizophrenia and alcoholism with the associated impairment of functioning as the magnifier. Bowen designed an enriched environment there, one that without external interference in the intense relationship left room for amelioration and could promote growth. He thought the family itself had innate strengths (if there was environmental support).

Each new observation refreshed Bowen's curiosity and inquisitiveness. According to Thomas Kuhn, after a discovery a person functions in a different world (Kuhn, 1970, 121). This was true for Bowen at Menninger after his literature search, again during his regression work, and definitely after his revelation of the family as a single organism. Bowen had been including families in his thinking from 1946, his first year of residency. That Bowen knew, at some level of awareness, he was on to a different theory at the outset of the research is supported by how he approached seemingly unsolvable questions such as transferring anxiety. He inductively recognized that the involvement of the staff was the spillover from the intensity of the mother–daughter relationship. To keep the upset between the mother and daughter required another family member.

This awareness is clear in the documents written up on the out-patient family he saw before starting the NIMH project and in Bowen's insight that "One does not create a new theory from errors about the old" (Bowen, n.d.a). How Bowen set up the ward at NIMH supports his assertion that he had already formed the basis of his theory by the time he arrived there. He noted it, years later, when it was safe to say so—when his reputation was intact and his ongoing work was protected.

> When people call me "intuitive, it is mostly a sign they go along with the former way of thinking …. It was not intuitive. I was merely trying to stay consistent with an orientation that sees the human as an integral part of all life on planet Earth. There is another discrepancy that is unwitting, that undermines my lifelong effort. **When I started the research at NIMH, I shopped around for the kind of symptom that would lend itself to 'research on human beings' that the institution could tolerate. It was at a conceptual expense to me, but families with 'hard core schizophrenia' did lend themselves to the requirement of the institution. I was working at 'getting at' the total human phenomenon thru an entrée that was schizophrenia.** [Emphasis added] I have written about that over and over but few hear it. Even during the NIMH years, I was working with non-schizophrenic people. When my research commitment to NIMH was over, I quickly moved to an orientation that involved all humans …."
>
> *Bowen, 1984*

The possibility that the family was a natural system, shaped by evolution, came from his explorations at Menninger. The idea lacked clinical evidence with actual family members. Experimenting with choices for his NIMH research environment produced the clinical evidence.

> The research was NOT directed at schizophrenia, as most believe, including too many in the Family Center itself. Schizophrenic people were chosen because they have the RIGHT kinds of families, and not because they are severely impaired.
>
> *Bowen, 1989a*

Bowen's theoretical orientation was in place prior to arriving at NIMH. There, he could immediately design a research setting—a laboratory—for further exploration with real family members with the most serious of human problems. Then he could, as he had in the past, extend the theory to understand others with less serious problems. Bowen described this in a 1990 letter.

> My first interest in theory began in 1946, almost fifty years ago. It was designed to help Freudian theory make contact with the rest of medicine. The effort did not go as planned. It resulted in a completely different theory, with a new series of clinical findings that could not be seen with traditional Freudian Theory. The whole operation was shifted to NIMH in Maryland for a five-year clinical trial. It was amazingly successful. The new findings included family therapy and a whole series of other findings called Family Systems Theory and Therapy, or the Bowen theory.
>
> *Bowen, 1990a*

The materials also show a self-deprecating side to Bowen. As he looked back, he disclaimed his genius, saying the theory was "a budding mess of disconnected ideas" (Bowen, 1968). He had "enthusiasm & cockiness about having an ans[wer] to schiz[ophrenia] when went to NIH in 1954" (Bowen, 1977). By 1987, he countered with conviction that

> the infinite detail of a different theory was pretty well worked out in Topeka before the move to NIMH, (8 ½ yrs.). That the new theory was merely put into research operation after the move to NIMH in 1954. That Family Therapy and the national focus on Family were merely immediate by-products of the new theory [Underline in original].
>
> *Bowen, 1987b*

In other materials from the same period, Bowen distinguished the Topeka-era form of the theory as relating to the two-person (parent/child) system. His NIMH investigation moved to relating to all the family. Bowen's research plan was a logical next phase from what he did at Menninger. Hospitalizing two members of a family was acceptable to a research institution with the perception that mothers were causative to the symptom of schizophrenia. Hospitalizing an intact family in 1954 would have been extreme, well beyond radical. In draft notes of his reflections on his odyssey, he connected Menninger and NIMH:

> In some nine years developed completely different theory based on a different formula about science. Basic theory based on a medical background about science. It included 1) scientific facts from Freud, 2) evolution appeared more scientific than Freud did. One does not go from Freud to Darwin easily. They do NOT go together. No way to connect the two without some form of Systems Theory. Ludwig von Bertalanffy had already created General System

Theory. It did not go with science. Another systems theory came from technology and war, mathematics (Weiner). No way to connect industry with the human. [I] created <u>Natural System Theory</u> (underlines in original) to connect the human with Darwin. Freud was a two-person system—one therapist + one patient. After 9 years, moved across the nation—at age 40 (young family, cut in pay)—to put new theory to work with multiple people. 1954 Freud + Natural Systems + reproductive family. There emerged a new theory about the human family ....The process may have been inductive to connect Freud + Darwin—but it was simple deduction after that.

Bowen, n.<u>d.b</u>

Bowen's theory on human capacities within a relationship originated in the explorations, assumption testing and research endeavors at Menninger. The clinical family data in the first year at NIMH extended the data with conviction to a new systems theory. Bowen made no public announcement of this beyond the March 1955 quarterly report. Instead, Bowen's work awaited the work of other scientists across disciplines for a convergence of evidence.

Courage to continue into the unknown

To further study two family member's interdependence, he brought mothers and daughters into the hospital together in 1954. In psychiatry, the perspective of symbiotic relations in 1954 did not come from the human as a biological being. Bowen believed it worthwhile to study these relationships. He replaced id, ego and superego with his usage of differentiation from embryology where one cell divides into two. This division continues. Clusters of cells form and, even though cells are alike and are side by side, they have autonomy to perform distinct functions that contribute to the whole organism (Bowen, 1980, 8).

This is the model for a system. Bowen had questions about this until the end of his life. What governed the division of cells? What guided the development of diverse functions between these groupings of cells: the cell, the clusters of cells or the whole organism (Bowen, 1990b)? Many of these questions remain to be answered even today. But Bowen used the word symbiosis with a biological base.

Psychological symbiosis was regarded as having the same meaning on the psychological level as the biological use of the term, 'two organisms living in an intimate association which is advantageous to one or both forms and not harmful to either.' The term as used here would thus apply to the calm, normal, advantageous closeness between a mother and infant.

Bowen, 1995, 32

Bowen was at a crossroads in his first year at NIMH. His findings showed that a disturbed family member was a physical representation of upset in the family group. Yet, individual theory and psychotherapy were the standard and accepted. Practicality favored continuing with these methods (Bowen, 1965, 121). An individual orientation was both familiar and sanctioned while his theory building challenged his tolerance for risk, perhaps even his position at NIMH, and certainly finding his way in the unknown. Bowen's principles directed his choice. He went with the theory building.

By studying the symbiosis while leaving it intact in the mother–child exchange, the parent, not the therapist, was the one showing the interest. Over the course of the first year, this led to understanding the family as more than a collection of individuals and it produced the family-staff sessions in 1955, the first effort at family psychotherapy.

When Bowen changed and implemented the refined hypothesis with intact families using family therapy as the practice method, it was the next important step in developing family theory. He acted on what the facts said while being convinced he was in unfamiliar territory. His psychiatric training gave him doubts about using an unknown, untested treatment such as family psychotherapy.

It was in January 1956, with the first intact family—father, mother and two sibs, one impaired—that each parent sought individual psychotherapy. By then, the family-staff meeting had been operating for several months. A review of those meetings showed some family members in the mother–offspring pairs took up their issues only in private therapy. They progressed less than the ones willing to speak face to face in the meeting. Bowen asked this newest family to forego separate therapy for six months and to make use of the potential in the family meeting approach. The "fair trial" lasted until June 1956. This method was so effective that it ended any individual therapy.

Bowen now had a way forward. The family as a functioning unit meant he could use his research operations to explore a theory based in evolution. Bowen was using new descriptive words by 1956 such as "undifferentiated family ego mass," "family psychotherapy" and "interdependent triad." He integrated these words with corresponding attitudes for himself and staff on relating to and treatment of the larger family unit.

For his remaining four years at NIMH, Bowen established and pushed the boundaries of the theoretical premise regarding a family. Round-the-clock observations continued as well as refinements in the family-staff meetings. He met with outpatient families with troubles less severe than schizophrenia. Bowen found that asymptomatic families, those with simple neuroses, and those with more severe impairments, had the same patterns "so striking in schizophrenia" (Bowen, 1965, 120). Bowen was unlearning parts of his Freudian training and moving toward universality of the family approach. It was the application to those with less severe problems or to those without identified issues that extended the "family concept of schizophrenia into the family theory of emotional illness" (120). By 1965, the theory applied to the entire range of human adaptation from the best functioning to the most difficult (120).

Documenting the roots of family systems theory

Bowen's ability to recognize that he saw evolutionary patterns in the NIMH families gives credence to his claims that a systems theory was in place before going there. He documented it in writings after 1980. He implied it in his actions while he was at NIMH. His building process was a gradational construction of theoretical scaffolding for such a declaration. Gathering information, observation, developing understanding and setting up ways to investigate family patterns across various levels of human functioning served as that scaffold. This resonates with the process that William Whewell, a nineteenth-century English historian and philosopher of science, suggested with his concept of "colligation."[1] Perhaps the environment at NIMH, which had an expectation of discovery and the permission to do such radical research, was the important variable. The design using the actual parent showed it was fortuitous that Menninger declined Bowen's research proposal.

The idea of a threesome as the smallest emotional molecule is a systems idea. A primary relationship unit was three, not two. Freudian theory had made the significant advance to a two-person system. Bowen advanced that to three people. He did not leave the observations at the level of the dyad or at interlocking pathologies, and neither at the individual idea of accepting schizophrenia in the person, nor as a manifestation of the relationship. The gathering of data from the start shows that the potential of the family as an entity concept was in the project's design. Once Bowen understood the concept of the family as a unit, he could see it in families with less intense difficulties (Bowen, 1965, 120).

That Bowen could understand what he saw in the way he saw it speaks to his forward thrust to a different theoretical base. No longer was the effort to extend Freudian theory to explain what he saw. There was now a solid base to define a different understanding separate from Freud's. Could his contemporaries hear this back then? In what way could he convey these ideas so that others heard them?

The new understandings were original enough, in substantive basic ways, to consider them distinct from Freudian theory. Even without naming it a unique theory, Bowen could practice from its principles based on evolution and systems in nature. The question was how long could he continue this work at NIMH?

Bowen's fit within NIMH

Scientific advance, while seeming sequential, often overlaps itself. Exposing revolutionary new ideas publicly leads to critiques of acceptance or opposition on one point or another, making it difficult to keep the whole vista of ideas in mind. Often, ideas that bring a fundamental change in perspectives must sleep for a century or more before acceptance can occur. If the press for a changed perspective is crucial to sustain life, the time is shortened.

By focusing on theory building, Bowen's project oriented differently than did other efforts within the Clinical Center. The other projects at the time were interested in families and the etiology of schizophrenia. They broadly encompassed socioeconomic status, social class, and comparison studies of those with and without the diagnosis in the general population (Farreras, Hannaway, & Harden, 2004, 264). Cohen, the Director of Clinical Investigations, wanted the working climate to offer the cross fertilization of ideas that occurs with exposure to the ideas of others.

> There was a great deal of incidental, informal contact, from which I learned an enormous amount, and I think the same was true for many others. We had a dynamic interplay between clinical and basic scientists ... an open-minded atmosphere of intellectual curiosity and social responsibility.
>
> *245*

> One of the major thrusts, not only in the Adult psychiatry Branch but in the entire NIMH intramural program, was to promote contact, lively exchange, and mutual assistance among the various scientists concerned with psychiatric problems.
>
> *255*

Again, how does one stay focused on data gathering for theory building when just talking about it raises serious concerns among others? Near the end of his life, Bowen addressed the difficulties:

> When a new theory was created back in the 1940s, it used terms never previously used in theory. It was decided to watch the new theory to see if new ideas extended or decreased lag time .... The author created a completely new theory in the 1940s. It involved a move into science, replacement of most of Freudian theory with the broader concepts of evolution and Darwin, retention of a small amount of Freud's thinking and combining Freud and Darwin with a special kind of systems theory.
>
> *Bowen, 1990a*

Bowen's project in a societal context

The concerted global effort to adapt to life-threatening challenges to survival during World War II spawned scientific and technological changes long into the future. Bowen was part of and contributed significantly to that enduring legacy.

> In the quarter century after World War II, the discipline of the history of medicine underwent dramatic change fueled by the spectacular development of science and science-based industry. A lavishly funded research enterprise attracted widespread public interest and spawned a new field of scholarship: the history of science.
>
> *Mills, 2021*

Bowen's project started on the cusp of dramatic societal changes in mental illness treatment, shifting from talking as the primary therapeutic agent to drugs as the primary therapeutic agent. Using psychotropic drugs resulted in a decrease in hospitalizations and took hold as a primary treatment.

In the first decades of the 1950s, psychiatry used patient interviews as the dominant treatment. Psychotic patients received more extreme measures, such as shock therapy with electric current or insulin-induced hypoglycemia. In the most extreme cases, they were subject to psychosurgery. From 1936 to 1956, it's estimated that 60,000 lobotomies were done in the United States and Europe (Faria, 2013).

By the conclusion of the 1950s, the changes in treatment of psychoses were drastic and far-reaching. "By the second half of the decade, there had been a large change in perception, a paradigm shift, based on the observations that chemicals could alter the mind, and the last lobotomy was performed in 1960" (Farreras et al., 2004, 298).

This was the context during which those in charge at the Center solicited projects that would advance understanding of human problems. The Clinical Center's lenience with Bowen's project of hospitalizing family members challenged normal operating procedures. Bowen recognized this when applying for a grant to continue his research after leaving NIMH.

> Every hospital has a strict administrative structure based on the concept of the individual as the unit of illness. Hospital structure requires the designation 'patient' and that the illness in the individual be named with a diagnosis. It required considerable flexibility and working together with the Institute for this project to exist in the hospital environment for the Clinical Center.
>
> *Bowen, c. 1960*

Termination of Bowen's project

Several times, Bowen's project was in danger of replacement. In 1958, David Hamburg became the permanent chief of the Adult Psychiatry Branch. Getting familiar with projects, he assessed the yield of the various family studies. Hamburg evaluated Bowen's family project as falling short (Bowen, n.d.c). Negative evaluations of Bowen's NIMH project argued "your findings do not apply to schizophrenia alone, your research should be terminated and replaced by a study designed specifically for schizophrenia" (Kerr & Bowen, 1988, 367). Other researchers "had already turned out reams of papers which were very impressive" (Robert Cohen Interview with Catherine Rakow, 2002, 36:25). Bowen's project had not. Cohen says the opinion at the administrative level was "there was too much related to the family" (38:05).

Hamburg's assessment and the dearth of published papers (important for continued funding) led to the closure of Bowen's program there in December 1958. Bowen noted he did not base his research on diagnoses, and he went unrecognized at NIMH as originating a theory of human behavior. This lack of recognition served Bowen's purposes to the degree he was not relentlessly having to justify what he was exploring. But it also brought about the project's demise. He wrote of this in 1983, in a letter to neuroscientist Paul MacLean. MacLean was in Basic Research at NIMH at the same time Bowen was in Applied Research. Bowen "admired (him) from a distance" (Bowen, 1983). "The idea that human behavior could, in any way, be related to animal behavior, was enough for me to continue a hidden agenda, without broadcasting the idea to the administrators who funded" (Bowen, 1983).

Negative evaluations of Bowen's NIMH project argued that Bowen was in a breach of promise, his study went well beyond the study of schizophrenia into new theory building and that exceeded his contract to specifically study schizophrenia (Kerr & Bowen, 1988, 367). Cohen, Bowen's superior, noted that Hamburg, "wanted to bring in some more biological studies to the program .... He wasn't criticizing him, [being] negative—he felt the others were being more productive" (Robert Cohen Interview with Catherine Rakow, 2002, 38:15). Cohen's opinion was that Bowen's leaving NIMH advanced his work. "Murray, in a way, benefited, I think, from being pushed out of the job (42:25) .... I don't know whether Bowen ... would've flourished as much, if he had stayed at NIMH ..." (49:07).

The family project was chaotic, causing upset in the broader organization. Bowen was not producing papers for the professional world. There was not yet a tangible product from his research. As a recipient of federal funding, the Clinical Center indirectly expected him to contribute to a positive view of the NIMH. In 1957, others in the Adult Psychiatry Branch published four research papers. Although he presented at conferences during his research years there, the publication of Bowen's paper came in 1959, after he had already left NIMH.

Bowen had a useful and important relationship with Lewis Hill. Hill, in turn, was very important to Bowen's supervisor, Cohen. Cohen had expected his relationship with Bowen to be equally useful but was disappointed. "I personally was, I must confess, somewhat disappointed" (Robert Cohen Interview with Catherine Rakow, 2002, 43:36). The misperception that Bowen's project was studying schizophrenia, coupled with his lack of interest in getting his numbers up and using a control group, as suggested in a peer review (NIMH Peer Review Record, 1957), gave an impression of uncooperativeness. Finally, Bowen was peripheral to the nucleus of the in-group. While Bowen was transparent about his project, and open to others observing, the endless discussions of theory at Menninger taught him to be careful in how he used his time. Others perceived his reluctance to engage in such discussions of his thinking as an unwillingness to be a part of the group. One place this occurred was in the research committee meetings for the Genain family. Colloquialisms such as "loner," "not a team player" or "not one of us" come to mind. Talk of replacing his project began at this time. These most likely contributed to his project being replaced. Robert Cohen, who engaged Bowen to come to NIMH, reflected on this in 2002:

> Now one thing I would say about Dr. Bowen was that he kept his – he didn't participate in the group process as much as the others. That is, I don't think

that I ever knew as much about what he was thinking. Because his papers –
he kept careful notes, et cetera, but he didn't produce so many and didn't take
part in the discussions.

Robert Cohen Interview with Catherine Rakow, 2002, 35:20

The time Bowen spent discussing a different conceptualization, explaining,
defending and convincing peers through his recitation of observations and hypothesizing in a group reduced the time Bowen had to advance and explore the theory
through publishing. He presented enough to get by at NIMH but published less
than others there. A new director of the Center took a hard look at the program.
Bowen's family study project ended.

Acceptance of the theory

Bowen considered lag time in the theory's acceptance within psychiatry. He had his
own version of diminishing time; the further he got from his original research in
the 1940s and 1950s, the clearer he was in describing what he knew and when he
knew it. Bowen addressed this many times. "It is often easier to be objective about
something in the past than it is when one is emotionally involved in the process"
(Bowen, n.<u>d.d</u>).

"I believe it has to do with invisible culturally determined forces that are refrac-
tory to observation while one is in it" (Bowen, 1974).

"Man can know something intellectually a long time before he knows it as part
of his being" (Friesen, 2012).

A common question is when Bowen knew with conviction that he had a
different theory as opposed to when he said he had a different theory. Bowen did
not claim an alternative theory early on. He patiently gathered clinical supporting
data over 20 years. Claiming an original theory is a chancy venture for a conscientious researcher. Without supporting facts, he risks being labeled as an eccentric
crackpot. "By 1956, I knew within me that the theory contained the necessary
variables to become an accepted science in the future" (Kerr & Bowen, 1988, 351).
Bowen first published the six original concepts in the theory in 1966, 20 years after
beginning his odyssey.

Bowen's approach, developed in the eight years before going to NIMH, showed
he repeatedly checked for facts. He was not dependent on a relationship to instill
ideas or reliant on a group to hash out repeated observations. Being at NIMH was
a strategic position from which to run his research, extend it and add to the theoretical concepts.

> The professional environment was hostile to neutrally accepting. I had to give
> up social relationships to stay on course. Gradually I had a few others on the
> beachhead with me. How did I get my working answers to theoretical and
> therapeutic dilemmas? I accepted no answers from the professional estab-
> lishment, nor from my own 'emotionally' determined 'clinical judgment.'

Every major decision had to come from the working hypothesis and from the research.

Bowen, 1989b

Bowen himself described his study using human subjects and how he maintained the neutrality needed to follow the facts. He set the NIMH ward up to observe, at the biological level, family members living together in one place, the hospital. From November 1954 to December 1958, ten in-patient families provided observational data and an investigation of family psychotherapy. Eight out-patient families having up to ten children contributed to the flexibility and generalizability of the study. He saw 12 families for detailed evaluations for admission, providing supplemental data (Bowen, c. 1960). On leaving NIMH, Bowen joined Georgetown University as an adjunct professor. He spent the next seven years in further research and pulling facts together into concepts.

> A <u>CONCEPT</u> is a partial theory. I have used the term to designate a <u>bundle of facts</u> that appear to be part of an overall <u>THEORY</u>, to await more <u>CONCEPTS</u> that will go together- Said in another way, a <u>CONCEPT</u> is on its way toward becoming a <u>theory</u> [Capitals and underline in original].
>
> *Bowen, n.<u>d.e</u>*

Bowen theory moves into the future

Archival materials can support just about any answer to the question of when Bowen's theory originated. In a 1958 paper, when his project was under scrutiny, Bowen wrote that his new understanding is "neither a theory, nor a partial theory" (Bowen, 1958a) and made a clear statement of the constraint of swimming upstream against then current understandings explaining his struggle to operate in a new world while living in the old one.

> It has not been possible to run the clinical and the research operation in complete harmony with our theoretical position. Our own limitations, and the generally accepted view that it is a disease in the individual, makes it necessary to retain part of the individual orientation.
>
> *Bowen, 1958b*

Bowen lived and worked in a world without claiming a new orientation. He tried to keep in good standing at the Clinical Center and to maintain relations with other professionals. Those were necessities for survival. Yet that did not stop Bowen from shining a light on the astounding possibilities of his discovery.

> I believe that when man is finally capable of such thinking, (a way to account for all the diverse data about man with a unified concept of man) he will understand the lawful order between the cell and the psyche and he may even

understand the lawful order between the psyche and the entity we know as soul.

Bowen, 1957

It is not common knowledge that Bowen's theory originated at Menninger. That may be the evidence of lag time. Bowen's research at NIMH was unusual and never replicated, and the Aha! discovery of the missing important systems element outshone the earlier research. Bowen recognized the missing clinical piece because he had a theory base that opened his mind to seeing it.

> My life has been devoted to the discrepancy between traditional psychiatry and the rest of medicine. Topeka was 'Mecca' for the traditional, and open questions about Mecca were not popular there. My head stayed on course in spite of opposition. Eventually it produced a new way of thinking about human problems .... The previous decade of thinking quickly produced family therapy, the magical word 'systems', and a host of new ideas never previously seen. For me, it was evidence that human behavior would eventually become an accepted science .... My goal has been a simple one which is to advance the cause of civilization as much as possible during my lifetime.
>
> *Bowen, 1987c*

At NIMH, Bowen's view of the human as a natural system allowed him to see beyond the view that the mother–child relationship was primary. He saw them as a fragment of a larger entity. His explorations at Menninger did not allow for this. It took the "in vivo" experience to generate evidence of the intense interdependence and participation of other family members. Bowen's training was in Freudian theory, and his experimentation examined tenets of that theory. His participation in and study of the relationship process in the regression were compatible with what was in the literature, yet the relationship process was not well defined in Freud's theory (Bowen, 1971, 119). He searched for clues to extend individual theory, clues that could move toward a more expansive biological theory incorporating related ideas from evolution.

Bowen's theory was operating at the two-person and three-person level at Menninger, but he did not make the leap to the family as a unit there. Operating possibilities there were not expansive enough for that. The Menninger mandate was dedicated to proving Freudian theory. A new environment minus the encumbrance from the institution, plus a research project designed to pursue his interests, supported the ability to see the round-the-clock intensity of the family relationships spilling out of the twosome. Bowen credits the live-in, 24 hour-a-day observations at NIMH for enabling him to escape his observational blindness that existed at Menninger. His interest in pursuing an evolutionary base to human action, going as far back as his literature search, brought new perceptions. The environment was the laboratory. He compared this "to changing the lens of a microscope from the oil-immersion to a low-power lens" (Bowen, 1959, 25).

Once seen, the new relationship phenomenon was so forcefully present that it pervaded the entire operation. It was then possible to see the phenomenon clearly in concurrent work with out-patient families in which the phenomenon was less intense in its manifestations.

Bowen, 1965, 119

Bowen withheld proclaiming a new theory until the evidence was convincing. He had a baseline theory when he first hospitalized mothers and offspring. Then he had the theoretically grounded insight that the family was the emotional unit. Nothing learned after 1955 disputed this. When an editor told Bowen he was close enough to an original theory to consider publishing it, this confirmed that others were hearing the theory (Bowen, 1962, 2).

Bowen could have come to the family as a unit idea sooner if symbiosis had not been in the literature (Bowen, 1964, 3). As late as a few years before his death, major questions remained for him on the acceptance of his theory. A theory, because of its definition, is never complete. Bowen's intention was to reach toward the science of biology and living systems while shedding the familiar comfort of the social sciences. It was a lifelong exercise of searching for data of human functioning and for exceptions.

Bowen's search for links to facts describing and explaining the human and the family is there in the early clinical records. One can see in his myriad of letters the determination to write of his findings in ways that resonated with others or, at least, that generated thoughtfulness, not debate. His effort has crystallized that for me. In 1956, in a letter to the father of his first regression patient (described in Chapter 6), Bowen wrote,

I am unable to separate the psychoses from the neuroses. Many new theories are proposed for schizophrenia, as if schizophrenia is something separate and apart. Yet, we have only to look around us and see every gradation of problem from the most severely regressed schizophrenia to the mildest neurosis. Indeed, there is not one among us who does not have some kind of a mild version of symptoms believed to be more or less specific for schizophrenia.

Bowen, 1956, 2

Bowen wrote this when exploring a scale of functioning to further his own understanding of differentiation.

What I discussed in this book is a brief ten-year window into Bowen's lifelong endeavor to understand human behavior. His pursuit of facts remained open to further inquiry. Bowen gave his view on when the theory would move into the mainstream. "Viable contact will have been made when new discoveries in the accepted sciences can be used in the research of human behavior, and when the accepted sciences start using discoveries from research on human behavior" (Bowen, n.d.f).

Bowen never claimed his theory was completely scientific. He stated it was his effort to move toward science and that it was a slow course requiring repeated and exacting checking of facts. In the last year of his life, Bowen still had important questions.

> I began using the term 'differentiation' over 40 years ago. It came from embryology in which one cell divides into two autonomous cells, and two cells into four cells, and so on … in the beginning, one cell is one bit of protoplasm. After the division, one cell becomes a heart cell, another a bone cell, another a brain cell, another a muscle cell, and on and on. What governs the process? Is it physical, genetic, psychological, physiological, or something else?
>
> In the later 1940s, I introduced the concept of 'systems' to bridge the gap in which Darwin replaces Freud. Now people use 'systems' as if there is something magic in the mere use of a word. Psychology plays a background role, but I do not know whether a healthier organism influences process organism, or whether systems thinking influences it all.
>
> *Bowen, 1990c,*

What defines the acceptance of a theory? For a theory of the human to meet the standards of science, it must meet certain elements and add meaning to societal and cultural norms. It must be "consistent with nature … and coherent with evolution" (Brown & Papero, 2017, 15:30). Science is moving toward understanding the commonalities of living things. When another's observations are available in our highly electronically connected world, a cross fertilization occurs. This enriches the transition and the conceptual shifts that go with advancing understanding. George Makari, historian, psychiatrist and psychoanalyst has noted, "Simply put, change in our theories comes from new observations" (Makari, 2002, 259). And those changes must be broadly and consistently available for further searching.

It was Bowen's observations of patients' deterioration when visiting with their families, particularly mothers because they were the ones who visited, which prompted him to change his experiments at Menninger (Brown & Papero, 2017, 17:38). "For a change in theory to be more than eccentric, it must be accepted by a community of patients and practitioners who are convinced of the utility of the theory" (Makari, 2002, 260).

The residents trained at Georgetown University, the thousands of students who attended the Bowen Center training program or programs at the 17 other US centers and international locations, and the scientists in other disciplines who are implementing these ideas in their research meet these criteria. This has created a cadre of researchers to further inform the theory.

The transfer of new observations to various disciplines studying living things occurs more rapidly in today's world. New, more educated observations and speedy transfer of information support advancement of alternative theories. As understanding moves to rules common among living things and expands the knowledge of natural systems, theoretical problems move beyond social utility. They become the territory of multidisciplinary researchers who accept the "internal challenge to increase the scope and precision of the fit between existing theory and nature" (Kuhn, 1977, 119).

Bowen's explorations contributed to knowing his own beliefs by assessing what he saw and having a sense of the mechanisms to understand them. The historical records document who Bowen was and how he lived out what he believed.

The archival materials provide the road map that Bowen followed: Get clarity on what a scientific theory should contain; search for and gather data based on factors about the human that fit evolutionary theory; use clinical work for support; stay open to gaps in knowledge knowing the family is like other life forms; notice the restorative possibilities if family members are part of the healing method; be willing to venture into uncharted territory when replacing assumptions with facts in theory; and take a chance and apply new ideas that appear to fit without having supporting current knowledge or experience with it, such as family psychotherapy. Prevailing negative views on families constantly challenged Bowen's attitude, whether the family caused deficit problems in an offspring or family was used for singular determinations of psychiatric symptoms such as disease or genetics. He left open for further exploration how other variables, such as the expression of genes, interact with the theory. Curiosity directed him to find alternative ways to understand observable phenomena. Approaching from an evolutionary perspective, he worked toward extending the theory to all human activities, from the family on through to society. It was a shift from individual diagnoses to system dilemmas (Bowen, c. 1956).

Note

1 "Colligation, according to Whewell, is the mental operation of bringing together a number of empirical facts by 'superinducing' upon them a conception which unites the facts and renders them capable of being expressed by a general law" (Snyder, 2000).

References*

*Unless otherwise specified, the works of Murray Bowen included in this chapter are from the Murray Bowen Papers. 1951–2004. Located in: Modern Manuscripts Collection, History of Medicine Division, National Library of Medicine, Bethesda, MD. The Accession, Box and Folder information differs for each one and that is included here.]

Bowen, M. (n.d.a). [Draft. Honolulu paper]. (Acc. 2006-003, Box 7, Folder Family relationships in schizophrenia).

Bowen, M. (n.d.b). [Draft. Third Thursday lecture]. (Acc. 2007-012, Box 3, Folder Science and therapy).

Bowen, M. (n.d.c). [Letter to Paul MacLean]. (Acc. 2007-012, Box 2, Folder Special working papers).

Bowen, M. (n.d.d). [Letter]. (Acc. 2005-055, Box 5, Correspondence from & to clients/ patients).

Bowen, M. (n.d.e). [Draft]. (Acc. 2004-043, Box 3, Folder Guerin).

Bowen, M. (n.d.f). [Drafts of family group therapy]. (Acc. 2005-055, Box 2, Working papers on theory).

Bowen, M. (c. 1956). [Draft. A psychological concept of schizophrenia]. (Acc. 2007-073, Box 3, Folder Working papers from NIMH project re: Schizophrenia).

Bowen, M. (1956). [A study of family relationships in schizophrenia and the therapeutic response when the family group is treated in a specific therapeutic setting, March 26, 1956]. (Acc. 2006-003, Box 4, Folder 3-E project).

Bowen, M. (1957). [Draft. A family concept of schizophrenia]. (Acc. 2006-003, Box 4, Folder Chapter drafts).

Bowen, M. (1958a). [Draft]. (Acc. 2006-003, Box 4, Chapter—A family concept of schizophrenia).

Bowen, M. (1958b). [Draft]. (Acc. 2006-003, Box 4, Chapter—A family concept of schizophrenia).

Bowen, M. (1959). Family relationships in schizophrenia. Chapter 3 in *Family therapy in clinical practice*. New York, NY: Jason Aronson, 1978.

Bowen, M. (c. 1960). [A description of a family research project for the study and treatment of schizophrenic patients and their families]. (2014-034, Box 6, Folder Project prospectus).

Bowen, M. (1962). [Draft]. (Acc. 2007-012, Box 3, Misc. correspondence).

Bowen, M. (1964). [Draft, "The Family as a psychological unit", Presentation to GAP, April 20, 1964]. (Acc. 2009-013, Box 6, Folder GAP meeting. April).

Bowen, M. (1965). Family psychotherapy with schizophrenia in hospital and private practice. Chapter 8 in *Family therapy in clinical practice*. New York, NY: Jason Aronson, 1978.

Bowen, M. (1968). [Draft of letter, No salutation.] Copy in possession of Bowen family, Williamsburg, VA.

Bowen, M. (1971). The use of family theory in clinical practice. Chapter 9 in *Family therapy in clinical practice*. New York: Jason Aronson, 1978.

Bowen, M. (1974). [Response to inquiry about practitioners on West Coast on October 18, 1974]. (Acc. 2005-055, Box 5, Folder Professionals requesting M. Bowen consult, advise, or refer).

Bowen, M. (1976). An interview with Murray Bowen. Chapter 17 in *Family therapy in clinical practice*. New York, NY: Jason Aronson, 1978.

Bowen, M. (1977). [Draft notes for presentation], (Acc. 2004-013, Box 3, Symposium 1977).

Bowen, M. (1980). Simon, R., "A network genealogy: A history of family therapy in the Baltimore-Washington corridor, Part III—The protoplasmic world: Conclusion of an interview with Murray Bowen. *Family Therapy Practice Network Newsletter, IV*(1), 1–10.

Bowen, M. (1983). [Letter to Paul MacLean on October 14, 1983]. (Acc. 2004-013, Box 3, Folder 20th Symposium, October. 15–16, 1983).

Bowen, M. (1984). [Letter to a conference organizer, June 1984]. (Acc. 2003-044, Box 5, Folder Name redacted, Chicago, IL, June 14–16, 1984).

Bowen, M. (c. 1986). [Letter]. (2007-012, Box 2, Folder Working papers).

Bowen, M. (1986). [Draft notes for chapter in Kerr/Bowen 1988 book]. (Acc. 2007-012, Box 2, Folder Working papers).

Bowen, M. (1987a). [Symposium, November 8, 1987]. (Acc. 2004-013, Box 3, Folder Symposium-Nov. 7-8, 1987).

Bowen, M. (1987b). [Letter to faculty, July 1, 1987]. (Acc. 2007-012, Box 4, Folder Working papers).

Bowen, M. (1987c). [Letter on July 31, 1987]. (2007-012, Box 2, Death & funerals).

Bowen, M. (1988). [Letter to a student in the post-graduate training program at the Georgetown Family Center on October 8, 1988]. (Acc. 2004-043, Box 5, Folder Letters from Dr. Bowen).

Bowen, M. (1989a). [Letter to a student in the post-graduate program at Georgetown Family Center on April 23, 1989]. (Acc. 2007-012, Box 1, Folder Interesting letters).

Bowen, M. (1989b). [Notes]. (Acc. 2007-012, Box 1, Folder Georgetown family center).

Bowen, M. (1990a). [Memo to adjunct professor, Georgetown university law center, July 6, 1990]. (Acc. 2007-012, Box 3, U. Georgetown med. sc.).

Bowen, M. (1990b, April 19). [Letter to a former resident at Menninger]. (Acc. 2003-044, Box 6, Folder Green Bay, Wisc.).

Bowen, M. (1990c). [Letter on April 19, 1990]. (Acc. 2006-003, Box 2, Folder Correspondence materials).

Bowen, M. (1995, Spring/Summer). A psychological formulation of schizophrenia. *Family Systems Journal, 2*(1), 17–47.

Brown, J., & Papero, D. (Hosts). (2017, November 8). Brown Podcast Bowen's World War 2 experience and early psychiatry training (No, 2). In *The life and times of Dr. Murray Bowen*. The Family Systems Institute. www.thefsi.com.au/conference-recordings_rashed/podcasts/.

Faria, M. A. (2013). Violence, mental illness, and the brain – A brief history of psychosurgery: Part 1 – From trephination to lobotomy. *Surgical Neurology International, 4*, 49. https://doi.org/10.4103/2152-7806.110146.

Farreras, I., Hannaway, C., & Harden, V. (Eds.). (2004). *Mind, brain, body and behavior: Foundations of neuroscience and behavioral research at the National Institutes of Health*. Amsterdam: IOS Press.

Friesen, P. (2012). Quote from Bowen in her presentation in Pittsburgh, PA., June 2, 2012.

Kerr, M., & Bowen, M. (1988). Epilogue, "An odyssey toward science". In M. Kerr & M. Bowen (Eds.), *Family evaluation*. New York: W.W. Norton and Company.

Kuhn, T. (1970). *The structure of scientific revolutions* (2nd ed.). Chicago, IL: The University of Chicago Press.

Kuhn, T. (1977). *The essential tension: Selected studies in scientific tradition and Change*. Chicago, IL: University of Chicago Press.

Makari, G. J. (2002). Change in psychoanalysis: Science, practice, and the sociology of knowledge. In P. Fonagy, R. Michels, & J. Sandler (Eds.), *Changing ideas in a changing world: The revolution in psychoanalysis – Essays in honour of Arnold Cooper* (pp. 255–262). London: Karnac Books.

Mills, E. (2021). Circulating now newsletter. In *So what's new in the past? The multiple meanings of medical history*. National Library of Medicine. Retrieved September 18, 2021, from: www.nlm.nih.gov/exhibition/so-whats-new/index.html.

NIMH Peer Review Record. (1957). [Memo to Dr. Bowen from peer reviewers, December 6, 1957]. (Acc. 2006-003, Box 4, Folder 3-E project).

Robert Cohen Interview with Catherine Rakow. (2002, June 12). *The Murray Bowen archives project*. Retrieved September 2017, from: http://murraybowenarchives.org/interviews/robert-cohen/

Snyder, L. J. (2000, December 23). William Whewell. In E. N. Zalta (Ed.), *Stanford encyclopedia of philosophy*. Retrieved July 28, 2021, from: https://plato.stanford.edu/entries/whewell/.

Westley, F., Zimmerman, B., & Patton, M. Q. (2006). *Getting to maybe*. Toronto: Random House of Canada.

INDEX

For Product Safety Concerns and Information please contact our EU
representative GPSR@taylorandfrancis.com
Taylor & Francis Verlag GmbH, Kaufingerstraße 24, 80331 München, Germany

www.ingramcontent.com/pod-product-compliance
Lightning Source LLC
Chambersburg PA
CBHW052121230326
41598CB00080B/3932

9 780367 461546